The Electric Industry *in Transition*

published by
Public Utilities Reports, Inc. and
New York State Energy Research
and Development Authority
1994

© Public Utilities Reports, Inc. and New York State Energy Research and Development Authority, 1994

All rights reserved. No part of this publication may be reproduced, stored in a retrieval system, or transmitted in any form or by any means, electronic, mechanical, photocopying, recording, or otherwise, without the prior written permission of the publisher.

This publication is designed to provide accurate and authoritative information in regard to the subject matter covered. It is sold with the understanding that the publisher is not engaged in rendering legal, accounting, or other professional service. If legal advice or other expert assistance is required, the services of a competent professional person should be sought. *(From a Declaration of Principles jointly adopted by a Committee of the American Bar Association and a Committee of Publishers.)*

First Printing, December 1994

Library of Congress Catalog Card No. 94-69987
ISBN 0-910325-58-8
Printed in the United States of America

NOTICE

This report was prepared by the New York State Energy Research and Development Authority, with cofunding from the Niagara Mohawk Power Corporation. Neither NYSERDA, NMPC, or any person acting on behalf of NYSERDA or NMPC: a) makes any warranty, expressed or implied, with respect to the use of any information, apparatus, method, or process disclosed in the publication or guarantees that such use may not infringe privately owned rights; b) assumes any liabilities with respect to the use of, or for damages resulting from the use of, any information, apparatus, method, or process disclosed in this publication.

TABLE OF CONTENTS

The Electric Industry in Transition

Foreword ix
William E. Davis
Chairman of the Board and Chief Executive Officer
Niagara Mohawk Power Corporation

Preface xiii

Introduction xv
Francis J. Murray, Jr.
Commissioner of Energy
New York State Energy Office
New York State Energy Research and Development Authority

Part I

Introduction
Competition in the Electric Industry:
Is it Inevitable and/or Desirable? 3
Marsha L. Walton
Associate Project Manager
New York State Energy Research and Development Authority

Chapter 1
"That Memorial Needs Some Soldiers"
and Other Governmental Approaches to Increased
Electric Utility Competition 7
Peter A. Bradford
Chairman
New York State Public Service Commission

Chapter 2
Competition in the Electric Industry: An Unguided Missile? 15
David Freeman
President and Chief Executive Officer
New York Power Authority

Chapter 3
Competition in the Electric Industry Is Inevitable and Desirable 21
Alfred E. Kahn
Special Consultant
National Economic Research Associates, Inc.

Part II

Introduction
What Does the Competitive Market Look Like
and How Will the Industry Work? 35
William J. LeBlanc
Director, Strategic Marketing
Barakat & Chamberlin, Inc.

Chapter 4
A Competitive Electricity Market: Lessons from the UK? 39
Graham H. Hadley
Executive Director and Managing Director
of International Business Development
National Power PLC

Chapter 5
The Shape of Things to Come:
A Competitive Market in the Electric Industry 51
B. Jeanine Hull
Vice President, Environmental and Regulatory Affairs
LG&E Power, Inc.

Chapter 6
Designing an Efficient, Competitive Electric
Transmission Policy and Access Market Model..................................57
Dr. William W. Hogan
Thornton Bradshaw Professor of Public Policy and Management
John F. Kennedy School of Government, Harvard University

Part III

Introduction
What are the Costs and Benefits of Market Efficiency
and to Whom? 67
Charles R. Guinn
Deputy Commissioner for Policy Analysis and Planning
New York State Energy Office

Table of Contents

Chapter 7
The Political Economy of Retail Wheeling,
or How to Not Re-Fight the Last War 71
Armond Cohen
Senior Attorney and Energy Project Director
Conservation Law Foundation

Chapter 8
Progressive Choice: The Customer as Regulator 93
Philip R. O'Connor, Ph.D.
Managing Director
Palmer Bellevue/Coopers & Lybrand
Terrence L. Barnich, Craig M. Clausen

Chapter 9
The Electric Industry in Transition 115
P. Chrisman Iribe
Executive Vice President
U.S. Generating Company

Part IV

Introduction
The Role of Regulators in Managing the Transition
to Competitive Power Markets 125
James W. Brew
Assistant Counsel
New York State Public Service Commission

Chapter 10
The Role of the Regulator in an Increasingly
Competitive Market: The Smaller the Better? 129
Leonard S. Hyman
Consultant and former First Vice President of Merrill Lynch,
Pierce, Fenner and Smith, Inc.

Chapter 11
The Role of the Regulator in a Restructured Electric Industry 141
Dr. John A. Anderson
Executive Director
Electricity Consumers Resource Council

v

Chapter 12
Protecting the Public Interest in the Transition to Competition
in New York Industries 151
Dr. Mark N. Cooper
Director of Research
Consumer Federation of America

Part V

Introduction
Transition Costs with Increasing Competition 169
Michael J. Kelleher
Director, Economic Research and Forecasting
Niagara Mohawk Power Corporation

Chapter 13
Methods for Dealing with Transition Costs for the
Electric Utility Industry 171
Dr. Theresa A. Flaim
Vice President, Corporate Strategic Planning
Niagara Mohawk Power Corporation

Chapter 14
Stranded Investments Costs:
Desirable and Less Desirable Solutions 185
Dr. Charles G. Stalon
Consultant

Chapter 15
Preparing for the Inevitable: The Nationalization of the
U.S. Nuclear Industry in a Competitive Electricity Market...............199
Dr. Edward P. Kahn
Economist
Lawrence Berkeley Laboratory

Part VI

Introduction
The Role of Technology in Serving the Customer's Needs 215
Arnold R. Adler, P.E.
ASME Legislative Fellow
NYS Legislative Commission on Science and Technology

Table of Contents

Chapter 16
The Role of IRP and Conservation in Electric Utility Transition 219
Frank Kreith, Doc.Sc., P.E.
American Society of Mechanical Engineers
Legislative Fellow, National Conference for State Legislatures

Chapter 17
Generation Technologies through the Year 2005...............................237
Anthony F. Armor
Director, Fossil Power Plants
Electric Power Research Institute

Chapter 18
Business and Technology Change towards the Year 2005:
Redefining the Electricity Industry's Customer Markets,
Products, and Services 253
Edward M. Smith
Chief Executive Officer/Director
Perot Systems Europe (Energy Services) Ltd.

Appendix I
Biographies of Authors and Moderators...273

Foreword

William E. Davis
Chairman of the Board and Chief Executive Officer
Niagara Mohawk Power Corporation

Competition in the electricity business is increasing, and the pace is quickening. Meanwhile, the Northeast has an oversupply of electricity that has driven down wholesale prices. At the same time, New York State's retail electricity prices have been rising rapidly, and are well above the national average. In order to compete, high-cost utilities are going to have to take drastic measures to bring their costs down. Some of the drastic measures being proposed, however, do not address the causes of the utility price increases, and if adopted would cause unnecessary economic harm.

Niagara Mohawk's average cost of generation is on the order of 5 cents a kilowatt-hour, including all fixed and variable costs of internal production, purchased power and property taxes. The wholesale market cost of power in New York State is now about 2.5 cents a kilowatt-hour. Some people look at the differential and reach two conclusions: first, that big cost reductions can be achieved for customers through retail wheeling and municipalization, and second, that utilities ought to write down assets to lower their costs to the 2.5 cents range.

Neither conclusion stands up to analysis. The bulk of the "gap" between the market price and Niagara Mohawk's cost of generation is attributable to circumstances that have little to do with the cost of our generating capacity, so a major writedown would be a solution largely unrelated to the cause. The most significant of those circumstances is the requirement to purchase power from unregulated generators. This year more than a quarter of our electric revenues, nearly $1 billion, will go to unregulated generators. We estimate that more than $350 million represents overpayment.

Other contributors to the perceived "gap" between wholesale price and utility costs in New York are taxes, forecast error, and current market structure. Utilities are the third largest tax collector in New York State, behind only the state itself and New York City. More than most states, New York has fastened on utilities as a surreptitious way of raising revenues. As a result, utility taxes in the state are more than twice the national average.

The Electric Industry in Transition

Forecast error caused unnecessary plants to be constructed in the Northeast. A lot of people were just wrong about where loads were going, about the cost of fossil fuels, and about the contribution of new technology.

The current wholesale price of electricity on the open market wouldn't hold up long if the marketplace was competitively restructured. Because of the relative oversupply of electricity, the Northeast at present is a strong buyer's market and, on the margin, generators will sell very cheaply, but no one can provide high-quality service for any length of time in a competitive market at these prices. If the market is completely restructured I suspect the market price will be considerably above current wholesale prices.

Of course our own costs, especially the nuclear costs that many advocates of writedowns point to, do contribute to the "gap" between the wholesale market price and our average cost of generation, but our total nuclear revenue requirements are about $500 million a year, while our overpayments to unregulated generators are about $350 million (Total payments to unregulated generators are around a billion.). Even a major writedown of our nuclear assets would not take us close to the current wholesale price level. The real problem is surplus capacity, not our nuclear costs.

To reemphasize that point, the major contributor to Niagara Mohawk's cost structure and the biggest barrier to equitable competition is not the company's nuclear operation, as many people believe, but a variety of utility mandates that have developed over decades, and which for Niagara Mohawk are principally excessive payments to unregulated generators for unneeded capacity and taxes. Since these costs were incurred under a regulatory system that has historically reimbursed utilities and their shareholders for investments that are prudently incurred, a transition to the competitive marketplace that is predicated on asset writedowns is simply unfair. There is little question of management prudence when costs are mandated by law.

The costs also include what Assistant Energy Secretary Susan Tierney has aptly called "stranded commitments"—those utility mandates that might not be supportable under more intense competition. These mandates include the universal obligation to serve, often at rates below cost; environmental requirements beyond those required of competitors;

Foreword

cross-subsidized financing of a variety of services; and social programs such as low-income assistance. Collectively they have a significant impact on costs.

An added regulatory challenge, particularly in a future that includes inter-utility competition, is that these costs vary widely from state to state. Regulators in the Northeast and on the West Coast have shown a greater tendency to impose societal costs on utilities than, for instance, in the South.

The costs of the transition to the competitive market will be very large. Estimates range from $100 billion to $500 billion, compared to total shareholder equity in investor-owned utilities of about $180 billion. Those costs are asymmetrically distributed across regions and among utilities. Some utilities do not have any stranded asset exposure and others have exposure of two to three times shareholder equity. An immediate jump to a competitive marketplace that includes retail wheeling—or a mass wave of municipalization that would have the same impact—could involve enormous redistribution of wealth among regions. The utilities that have the highest rates are generally going to have the highest stranded asset exposure, and would suffer the most serious consequences during a sudden shift to wide-open competition.

So how should we proceed with the transition? Certainly, it is going to occur in some form—given the degree to which competition has already entered the electricity business. Therefore, where we are headed is not as pressing a question as how we get there. An even more compelling question is who pays the transition costs?

In Niagara Mohawk's judgment, a rapid unguided transition will visit most of the harm on utility shareholders and small, mainly residential, customers. Only a comprehensive, carefully planned transition can avoid damaging impacts. We advocate a balanced approach involving reduction of some utility costs and redistribution of others. Reducing costs would allow utilities to use the savings gained to accelerate the depreciation of uneconomic assets, while at the same time minimizing rate increases.

Those redistributing utility costs should seek to share the pain, and the benefits, of the transition more equally. Redistributing too large a share of the costs to captive customers is not only unfair but in many jurisdictions a political impossibility. And there are competitive limits to rates even for customers now considered captive, given the ability of cities, towns, and other political jurisdictions to form municipal utilities and gain access to generation at depressed market prices through open transmission access at the wholesale level.

Utilities, too, should bear their share of the costs. A balanced transition plan might include some writedown of uneconomic assets, if it is fair and based on market valuation and if it is accompanied by a write-up of transmission and distribution assets that would have increased value in a competitive market. Our proposed approach to the transition would spread the costs and benefits among all competitors and customers. It would reduce burdens that have little to do with our business—like taxes.

However, we do advocate continuation of some environment, energy efficiency and social programs that serve the needs of society and are tied to the provision of electric services. Their costs must, however, either be reduced through movement to more market-based approaches or spread more equitably among all competitors.

It is also important that those who benefit most directly from competition pay their fair share of the transition costs through mechanisms such as exit and access fees imposed on customers who leave or return to the system.

We also believe the transition should proceed at a measured pace, so that Niagara Mohawk and other utilities can respond to emerging competition without having to reallocate existing costs abruptly.

Clearly the transition will occur whether or not the utility industry, its competitors and its regulators work to guide it. It will take all of the talents of all the interested parties, such as the capable individuals who assembled at this conference, to effect a solution that preserves the best of traditional utility service and at the same time realizes the benefits of a more competitive market.

Preface

This book is a collection of papers presented at a conference on the Electric Industry in Transition, June 14–15, 1994, in Albany, New York. Five hundred attendees representing public and private interest groups, State and Federal agencies, utilities, and independent power producers heard industry experts discuss the effects of a changing electric market on the electric industry.

After the conference was held, the New York State (NYS) Public Service Commission convened Phase II of its Competitive Opportunities Proceeding[1] and urged interested parties to work collaboratively to identify some principles to guide the transition to a more competitive market in New York State.

We hope the conference brought us closer to developing realistic policies and a regulatory framework to help the electric industry in New York reposition itself for a successful future.

This steering committee coordinated the conference:

- Arnold R. Adler, NYS Legislative Commission on Science & Technology
- James W. Brew, NYS Department of Public Service
- James E. Cuccaro, Orange and Rockland Utilities, Inc.
- Theresa A. Flaim, Niagara Mohawk Power Corporation
- Charles R. Guinn, NYS Energy Office
- Leonard S. Hyman, Independent Consultant and former First Vice President, Merrill Lynch, Pierce, Fenner and Smith, Inc.
- Edward P. Kahn, Lawrence Berkeley Laboratory
- Michael J. Kelleher, Niagara Mohawk Power corporation
- William J. LeBlanc, Barakat & Chamberlin, Inc.
- T. Davetta Montgomery, NYS Energy Research and Development Authority
- Marsha L. Walton, NYS Energy Research and Development Authority

---

[1] Proceeding on Motion of the Commission to Address Competitive Opportunities Available to Customers of Electric and Gas Services and to Develop Criteria for Utility Responses, Case 93-M-0229.

The Electric Industry in Transition

The New York State Energy Research and Development Authority and the Niagara Mohawk Power Corporation co-sponsored both the conference and this book.

INTRODUCTION

Francis J. Murray, Jr.
Chairman
New York State Energy Research and Development Authority

Bob Dylan could have been envisioning the electric utility industry today when he first sang, decades ago, "the times, they are a changing." For if there is one word that can most accurately describe the electric utility industry today, in New York and the United States, it is "change."

Change is rarely welcomed. Human nature being what it is, we are invariably more comfortable with the familiar, even as we acknowledge its imperfections and occasional inequities. Change is also difficult. Change is fraught with uncertainty and unpredictability.

Why should the electric industry react any differently? If anything, in the electric utility industry, this sense of discomfort, confusion and anxiety is compounded by the complexities of the industry itself, by the Byzantine world of utility regulation, and the almost mystical powers impugned to terms such as "the regulatory compact," "marginal cost pricing," and "environmental externalities."

At times, the temptation is powerful to emulate the ostrich—with its head in the sand—and to chant to oneself the refrain "this too shall pass." But that would be a mistake. For it won't pass. The one certainty is that change is inevitable. We cannot ignore such change, except at our own peril. We need to talk about that change—what it means not only for the electric utility industry, but also for all the other actors and interests that have a stake in the outcome of this evolution of change.

In New York, like many other states, the electric industry is struggling to adapt to intensifying price competition. Over the last decade or so, New York's electric industry has evolved from a structure characterized by large-scale, centrally planned utility generation stations to an industry where generation is more dispersed, the supplies more diversified, and the actors more varied. The emergence of a vibrant, economically viable independent power industry has been a particularly important factor in transforming the nature of New York's electric utility system. Consequently, today in New York the electric utility industry consists of a shifting mix of rate-based generation, cost-based wholesale power agreements, avoided cost-based purchase power contracts, and competitively priced power.

We have rising retail electricity rates. We have a substantial current surplus of electricity in New York with marginal costs considerably below average costs. Electric customers, initially primarily large industrial and commercial customers but increasingly other classes of electric customers, are demanding greater choice. The emergence of new, more efficient and cleaner technologies provide potentially new options for electric customers. Combined with passage of the National Energy Policy Act of 1992, which some argue offers new possibilities for state regulators to promote increased competition in the electric industry, this confluence of events is driving the emerging competition in the electric industry.

How inevitable is this new era of competition? How far should it go? For some, it has become almost an article of faith, indeed a theology, that the marketplace is better qualified to make decisions about energy investment and power purchases than either government or a heavily regulated, centrally managed utility. If somehow we could just release the competitive instincts and managerial ingenuity of an unregulated, or at least considerably less regulated marketplace, so the argument goes, then we will see energy costs lowered for many electric customers.

If you don't embrace this theology, however, as many environmentalists and consumer advocates do not, or if you are primarily concerned about the long-term implications of short-term actions, you may have a very different view of what increased competition, especially at the retail level, means for such important issues as environmental quality, demand side management programs, residential customer rates, low-income energy assistance programs, investment in renewable energy and other emerging technologies.

Setting aside the specific impact of increased competition on the electric utility industry itself, what does this "new era of competition" mean for New York State? For job creation and economic growth? For environmental quality? For all ratepayers? For system reliability, energy efficiency, and diversity of supplies?

To begin to address some of these questions, the New York State Energy Research and Development Authority and Niagara Mohawk convened in Albany a conference on "The Electric Industry in Transition." This

Introduction

book of papers is a product of that conference. It reflects a wealth of policy perspectives on the potential strategies and outcomes of a more competitive electric market. It includes the views of some of the most talented and respected individuals in both the public and private sector today who are engaged in what has become a national debate over the future of our electric utility industry.

For the electric industry, the change that competition is bringing presents both new opportunities and new risks. But government also is confronted with challenges. We too must change. That too is a theme in many of these papers. Most regulators recognize that the current regulatory system—which relies upon traditional rate-of-return regulation and which assumes a world of monopolies—is ill-suited to a new world of competition. The tougher question is what should we replace this system with. In this new world, the public sector must respond by developing a regulatory scheme that accommodates change, encourages flexibility, promotes cost-efficiency, and protects customers unable to avail themselves of new options.

While government cannot dictate change in this area, we can influence significantly the pace and form of change. Our objective is to promote policies that benefit all New Yorkers and that make New York a more attractive place to do business. Our primary goal will be to provide a safe, affordable and reliable supply of energy that is essential for economic growth and social prosperity. In an increasingly integrated economy, this goal cannot be realized in isolation from other compelling public policy objectives.

As we collectively confront the brave new world of competition, public officials have a special responsibility to remember the need to balance near and long-term interests. New York's challenge is to balance the need for secure and flexible energy services, a safe and clean environment, and a competitive and growing economy in a manner supportive of economic development. In addressing this challenge, we must be cautious about taking actions that might improve utilities' competitiveness in the short-term, but which work against realization of some very important, long-term policy goals, including energy security, environmental quality, technological innovation, energy efficiency, supply diversity, and sustained economic development.

The Electric Industry in Transition

As these papers demonstrate, there is no precise roadmap to guide our journey. Ultimately, how we respond to this tension—whether as a government or an industry—will reflect more deep-seated values. To the extent that these papers enlighten our way by contributing to our knowledge and wisdom, this conference will have been a success. To all who participated, especially our panelists, I express my deepest appreciation. We are in your debt.

Part I

The Electric Industry in Transition

▶

INTRODUCTION

Competition in the Electric Industry: Is it Inevitable and/or Desirable?

Marsha L. Walton
Associate Project Manager
New York State Energy Research and Development Authority

The New York Public Service Commission's decision to allow a major industrial customer in New York to purchase electricity directly from the cogenerator that supplies its steam and to require the cogenerator to compensate the franchise utility for a portion of its resulting lost sales in the form of an exit fee is an indication of growing competition in the electric industry in New York State and the pivotal role that regulators will play. In light of this decision and of California's and other states' consideration of direct access (retail wheeling), the question of competition's inevitability may sound naive. But inevitability implies there is no choice, and in each of the cases thus far, a conscious decision was made to encourage competition in electric markets.

Whether competition is desirable is then a moot point for those states that are deliberately encouraging competition, but the questions still remaining are why is competition considered desirable, and under what conditions? To the extent that competition forces utilities to reduce costs, resulting in lower rates for customers, it is desirable. Competition could force the utilities to tighten their belts and "race with whippets," to quote New York Public Service Commission Chairman Peter Bradford, or it could force them to cut critical corners to stay in the generation race. In attempts to cut costs utilities may delay the capital investments necessary to maintain their nuclear plants or underinvest in broader societal goals like research and development for clean air and sustainable energy. To the extent that competition encourages utilities to focus on cutting near-term costs and discourages long-term investments in reliability, a diverse fuel mix, integrated resource planning, and investments in renewable and indigenous resources, then the near-term rate reductions hold hidden costs that could result in more expensive rates in the long term. If competition encourages performance-based cost recovery and appropriately allocates risks between customers and energy providers, then competition would be desirable from the customers' perspective. However, if competition results in simply reallocating costs to core customers, then it would be undesirable from an equity perspective. If competition results in stranding utility investments[1] in plants that might be cost-effective over the operational lives of the plants but that cannot compete economically in the near-term with independent power producers' low costs, is competition desirable?

The societal implications of the electric industry moving from a regulated monopoly to a competitive market could be enormous. Under a regulated monopoly, utilities' investment decisions and recovery of expenditures in customer rates are closely scrutinized and approved only if they are deemed to be in the public's best interest or "prudent."

Is competition in the electric industry inevitable or desirable and the implications of these questions are addressed in the three papers in this chapter by a New York regulator, an economist and a public power provider. The authors unanimously agree there are desirable consequences of competition, but each qualifies his support of increased competition. The consensus is that in a competitive arena, the rules of the game will be all important in determining whether society as a whole is better or worse off.

New York Public Service Commission Chairman Peter Bradford is confident that increased competition will result in lower costs to customers and more choice in types of services offered. But at the same time he stresses the need for deciding the form that competition should take and the importance of sequencing the transition and establishing "unconditional imperatives...principles that we will steer by, no matter what." He makes the sobering observation that under the inevitability of inconsistencies among imperatives, it may not be possible to entertain more than one unconditional imperative at a time. David Freeman, president and chief executive officer of the New York Power Authority, is optimistic about the kinds of changes that competition will bring about in the electric industry so long as market competition considers the

---

[1] "Stranded investments" refer to costs and payments to nonutility generators that utilities will be unable to recover from base rates due to competitive pressures to reduce rates. Stranded investments also refer to utilities' capital assets that the utilities will be unable to continue operating because the operating costs exceed the market's marginal price of providing electricity.

Introduction: Competition in the Electric Industry

things that are important to society. He believes that the utilities' goal should be that of divestiture and separation of the generation, transmission, and distribution functions of the industry. Alfred Kahn, professor emeritus at Cornell University, former regulator, consultant to the industry, and champion of the market to allocate resources efficiently, describes the distortions of competition on an uneven playing field and proposes a "second-best economic solution" to permit recovery of the utilities' sunk costs while ensuring that competition is efficient.

Bradford opens the gate to competition in a measured way. He advocates establishing wholesale competition and considering revenue stability before considering direct access. Since 15 percent of New York's total electric bill goes to pay State and local taxes, he suggests that some degree of rate and demand-side management (DSM) stability could be achieved through tax reform. He acknowledges that most of the efficiency benefits of competition will be realized though wholesale competition, but realizes that other adjustments will be necessary to improve efficiencies at the transmission and distribution level. Although he recognizes that the downsizing that will result from increased competition will limit the industry's ability to pursue social agendas, he does not think that New York utilities' commitment to cost-effective energy efficiency investments should suffer as a result of competition.

Kahn, writing from his experiences deregulating the telecommunications, airline and trucking industries, provides a three-part explanation as to why competition in the electric industry is desirable and probably was inevitable. He claims it was probably inevitable because of the large discrepancy between the price that customers pay for electricity and the independent power producers' cost of producing it. At the same time, competition is desirable to promote a more efficient allocation of resources. The commercialization of smaller, less capital-intensive generating technologies contributed to the large discrepancy in prices, and the abrupt decline in demand for electricity exacerbated the utilities' high average costs relative to their competitors. Kahn also notes that the historical trends in inflation, skyrocketing fuel prices, and the abrupt decline in the growth of demand have all contributed to the utilities' high costs and uncompetitive position.

Kahn disagrees with Bradford and Freeman that the form and pace of competition can be controlled. He does not think that competition can be limited to wholesale markets. Kahn claims that there is a tendency for competition, once introduced, to become more and more pervasive even without price disparities. He predicts that every part of the electric industry will eventually be provided competitively. He points out that the utilities' obligation to serve will produce distortions in competition among rival suppliers, and that regulators must be prepared to treat all competitors equally.

Kahn advocates assessing a transmission charge to help utilities recover a portion of stranded investments and to support energy-efficiency and universal access. He proposes that utilities should be able to charge competitors "the proportionate contribution to stranded costs and social programs (including universality of service) that is embedded in their own rates for the business that they lose to those competitors, by recovering them in access, transmission and distribution charges," not only in order to recover stranded costs but also to ensure that the competition is on the basis of the relative efficiency of the contesting parties. Nonetheless, he notes that the recovery of sunk costs will increase rates above marginal costs and, therefore, will not be the first or best economic alternative. For this reason, he pragmatically advises utilities to get "as much of those (sunk) costs back as possible by writing down the book value of inadequately depreciated plants, while the getting is good."

Whereas Bradford and Kahn are more concerned with how, when, and where competition will manifest itself in the electric industry, Freeman is more intent on deciding where society as a whole wants to be in the next century and how the electric industry can help us get there. Freeman is critical of California's proposed direct access and claims that all the debate about retail wheeling is misplaced. It should instead be focused on developing "more cost-effective technologies that are environmentally sound and that meet the needs of society."

The three authors contributing papers to this section lay out some difficult tasks for policymakers and regulators in New York, who must adapt current regulatory practices to allow the market to work more efficiently and eliminate unnecessary costs to customers; build in the proper checks and balances to maintain the world-class integrity of New York's electric system; and, for reasons of efficiency and equity, ensure that all competitors equally support the development of renewable and energy-efficient resources, universality of access, and environmental protection. Otherwise, as Freeman aptly concludes "...lower prices today are a false bargain for tomorrow, and this industry has always looked out for tomorrow."

CHAPTER 1

"That Memorial Needs Some Soldiers" and Other Governmental Approaches to Increased Electric Utility Competition

Peter A. Bradford
Chairman
New York State Public Service Commission

Competition in the electric industry is both inevitable and desirable. The important questions have to do with the type and pace of competition, and also with first principles or what the philosopher Karl Jaspers called "unconditional imperatives." By this he meant—or at least I mean—the principles that we steer by, no matter what. All other principles will be pursued to the extent that they don't collide with an unconditional imperative.

For example, policies that take retail competition as an unconditional imperative may seek to protect stranded assets only to the extent that they don't compromise progress toward competition; whereas a policy that makes the recovery of stranded investment or rate stability for captive customers or the attainment of environmental goals its first principle will slow its pace toward competition to accomplish these ends. The important point is to realize that there is always an unconditional imperative at work, although it is often unstated and maybe something as general as "maximize consumer (or voter) happiness".

During comment periods such as the one now underway in California, one can for a time intend to make all the major imperatives unconditional, but in the end hard choices must be made—explicitly or implicitly.

Failure to make such ranking choices among inconsistent imperatives brings on a realization that not all of the promises to hold harmless can be kept. This realization in turn triggers a state of paranoia—a state whose capital we may for now call Sacramento—in which each party assumes that the promises protecting it are the ones to be broken.

The New York Public Safety Commission (PSC)—like 48 other state commissions—is presently content to play tortoise to the Californian hare. The most that I can do today—given pending proceedings and my own uncertainty—is offer some observations, including a few about contenders for the title of unconditional imperatives.

First, on the general subject of competition: where our society has been able to introduce competition into monopoly areas, we have had scant cause to regret it although the transitions have sometimes been much more wasteful and/or confusing than necessary.

◆ Total costs always fall; they never rise.

◆ Productivity always increases; it never declines.

◆ Customers as a group always wind up with more choices, never fewer.

◆ Technological advance and new products are always spurred, never retarded.

◆ Responsiveness to customers improves; it does not degrade.

◆ We never wish to restore regulated monopoly where it has been superseded by genuine competition.

Despite the validity of these generalizations, there are clearly downsides to downsizing. The ability to pursue social agendas—desirable or otherwise—is circumscribed. Those customers whose prices had been below their costs of service or who preferred not having to make choices are likely to be worse off. Responsiveness to customers who remain captive may decline even as pressure on their prices increases. Careless deregulation may free up monopoly predation where competitive checks are inadequate. The availability and/or lucrativeness of jobs in the former monopoly will decline even as the surviving senior executives sometimes announce bonuses whose justification in enhanced productivity may be slightly perceived or just plain slight.

Companies facing competition have to be more responsive to their customers but they have to be very cost conscious in dealing with everyone else, including taxing entities, suppliers, their workforce, their environmental commitments and their charities.

Chapter 1: That Memorial Needs Some Soldiers

Second, New York's draft State Energy Plan says that most of the benefits of competition can be had from whole sale competition. Our hearings on the Plan have shown also that most of the opposition to the plan focuses on retail competition.

This suggests the wisdom of preparing for wholesale competition first. However, if we are to rest on those oars, some other force is going to need to energize the transmission and distribution functions. Perhaps performance-based regulation will do it, but something must. When one compares the vigor of New York utilities' pursuit of gross receipts tax reform with the vigor of their encounters with independent power producers, one sees the difference between their reaction to potential competitors and their reaction to costs that flow through.

Similarly, New York utility service territories upstate look like the work of a Parker Brothers employee on an epic binge, circa 1930. They are nothing like the most cost-effective franchise lines for serving the utility customers.

Third, the issue of stranded assets in a competitive environment should not drive fundamental decisions about competition, but it does present vital questions of efficiency and equity. I don't think that a compelling case can be made for a historical regulatory compact that compels the recovery of every prudently invested dollar. As Irwin Stelzer has pointed out, for the last two decades utility executives and Wall Street analysts have been decrying the inability of regulators to keep their part of the bargain, so there can't be any investors left who bought stock in blind reliance on an impregnable compact.

Nevertheless, there are sensible reasons for permitting recovery of prudent investment even when it is retired early. And there are equitable reasons for making sure that utility investors do not suffer as a result of governmentally-imposed missions.

The prudent investment standard for recovery is the only one that gives utilities a basis for maximizing the efficiency of their operations on an ongoing basis. Any standard that tells a manager that more costly plant must be used or that competition must be suppressed if investment is to be recovered conveys a set of incentives that are not in New York's best long-term interest.

Neither does it seem defensible for the state or its subdivisions to stuff the private utilities with costs—even beyond the extent to which the utilities have stuffed themselves—and then lead them to a starting line to race with whippets—or even Power Authorities—without some opportunity for training and dieting. Of course, if one wants races soon, that training opportunity may temporarily take the form of handicapping rather than postponement.

Fourth, those who see future competition as a basis for eviscerating present energy efficiency programs should think again. The success of energy efficiency is a principal reason why I don't have to spend the 1990's as Fred Kahn had to spend the mid-1970's, keeping a weather eye on each OPEC meeting as he dealt with major siting board cases between rate cases and drowning in disputes about construction work in progress (CWIP), which seems now to be one of those dimly remembered New Deal acronyms.

It would, of course, be infinitely preferable for the federal government to reflect its concerns about security and the environment in prohibitions or in the prices of the relevant fuels so that the market instead of the regulators could value them appropriately. Since that day is not imminent, competition for electricity implies an advantage for fossil fuels underpriced in relation to their environmental and security impacts. Much resistance to competition rests on this circumstance.

And, of course, one of DSM's strengths is its flexibility. When the value of a saved kilowatt or kilowatt hour falls, these programs can be adjusted. The New York PSC is also moving to improve the greatest source of discontent with DSM by reducing nonparticipant rate impacts.

Our fundamental commitment to pursuing cost-effective efficiency is undiminished. While it is too early for final judgments, the results of some of the utility industrial DSM programs seem to rebut those who on the one hand asserted that energy audits were a waste of time for large, sophisticated customers and those who on the other said that such tailored programs would be no more than a pretext behind which industrial customers would evade DSM costs. The bottom line is that we are not implementing competition in ways that undermine cost-effective efficiency.

Fifth, our thinking will be much clearer if we reach some consensus definition of economic development, perhaps something in terms of

Chapter 1: That Memorial Needs Some Soldiers

growth in New York residents' disposable personal income. Right now everyone claims to embody economic development and everyone invokes it to lower his or her own costs regardless of the impact on those of all other consumers. The concept is clearly vital, but the term—as most parties toss it about in our proceedings today—has no useful meaning.

Finally, let me set forth some vague numbers before concluding with a precise generality.

The total electric bill paid to New York utilities, including the Power Authority, by retail customers at the end of 1992 was some $14.5 billion. Five years ago, it was $11.5 billion. Five years before that it was $10.9 billion. Of this sum, $1.3 billion (about 10%) went to independent power producers, $3.5 billion (about 25%) went to support the utilities' own generation (return on investment plus operating costs), $2.1 billion (15%) went to state and local taxes and $253 million (less than 2%) went to demand-side management.

To give you a sense of the dollar impacts of large power plants from different generations, $268 million went to Indian Point 2, $344 million to Indian Point 3, and $756 million to Nine Mile 2. Commencing in 1995, I would estimate the annual revenues paid to the Sithe/Independence 1000 MW facility to be in the $400 million range. The IPP and tax figures will also have increased substantially.

To put these numbers in another perspective, the total 1992 state electric bill divides into about $800 per New Yorker or $3200 per year for a family of four. Of course, the typical family annual electric bill is not $3200. That figure instead represents the family electric bill plus the electric cost component of all other goods and services. Even then, the number assumes some rough balance between imports and exports from New York State.

These numbers will suggest different conclusions to different interests. They seem at least to suggest that one could preserve a measure of rate and DSM stability through some modest tax reform.

However, these numbers are never viewed in the aggregate that you have just heard them. Instead, they are debated in a dozen or more separate proceedings. Bill Davis gives you a sample by describing a Niagara Mohawk profile accurately enough, but it is not one that would apply to the state as a whole or, for example, to the Long Island Lighting

The Electric Industry in Transition

Company (LILCO), which has fewer non-utility generators (NUGs) and a stranded asset once called Shoreham.

What is most frustrating about the present situation is the difficulty of getting all issues on the same table. Debates rage separately about rate levels, clean air goals, tax reductions, independent power producer contracts, stranded commitments of various sorts, utility plant retirements, economic development rates, Power Authority discounting, and the appropriate levels of utility demand-side management.

When these issues are addressed separately, the "solution" to any one problem tends to exacerbate others. Parties cannot make concessions in one proceeding because the concessions that they need in return are controlled by other parties in other proceedings before other decisionmakers.

All sides pledge themselves to efficiency, reliability, cleanliness and job growth. Then—in Lincoln's words—each invokes these terms against the other.

Supposing—just for argument's sake—the State of New York were to say that $16 to $17 billion per year is as much as we ought to be paying for electricity in any year between now and the year 2000. That is somewhat above what we pay now. The state could commit itself to doing its part through gross receipts tax reduction and perhaps through some limitation on other taxes as well as by limiting new clean air requirements and by facilitating access to inexpensive NYPA power or power that NYPA could purchase. NYPA could also cooperate in any savings that could be achieved in combined nuclear operations and perhaps in NUG buyouts.

Except at the extremes, regulation under such a cap would not focus much on earnings. Service quality would have to be maintained. Some rate design flexibility would be in order, but major rate shock or impact shifts would be unacceptable. Allowances for revenue growth resulting from economic growth would be made. Because revenue rather than rates would be capped, cost-effective energy efficiency would not be constrained. Utility efforts to enhance their productivity would produce savings that could be applied to the stranded investment problem.

By the same token, some discussion of trading off a level of early plant retirements for concessions from the NUG community might also be possible. This aspect of the discussion would have also to include a utility commitment to separate (and probably ultimately to sell off) generating

Chapter 1: That Memorial Needs Some Soldiers

assets in order to assure meaningful wholesale competition. Transmission assets would have to be operated as a coordinated pool reflective of short-run cost variances. Consideration of retail competition could be postponed at least until the expiration of the revenue cap in the year 2000. The rate flexibility necessary to meet competition would have to be achieved under the caps.

It seems to me that some effort to achieve a larger discussion forum—if not along the foregoing lines then along some other—is going to be necessary both in the state energy planning process and in our own PSC proceedings.

I have set this notion forth not because I think it is the right approach—I don't know that—but because I wanted to show you a somewhat different way of thinking about the issues that challenge us, a way that takes some type of total revenue stability as its unconditional imperative and pursues at least wholesale competition within that framework.

CHAPTER 2

Competition in the Electric Industry: An Unguided Missile?

David Freeman
President and Chief Executive Officer
New York Power Authority

Competition is a very vital, a very strong force, but those of us in the utility industry that have been in monopolies most of our lives are having a little difficulty warming up to this headlong mad dash towards competition. I happen to run a utility that is facing competition, and the upshot of it is that I spent all day yesterday, as I will spend all day tomorrow and the day after that, cutting costs. Competition is having a major impact on the New York Power Authority and the way we will do business in the future. We will reduce our cost of producing electricity by $70 million per year, or nearly 15 %.

The question we need to ask is, "Competition for what?" This is a very serious question. Successful enterprises must respond effectively to the market place. But it is my own experience, and I think history will tell you the same, that very rarely do we find successful firms that are simply slaves to the short-term market. This industry, over its history, and through these glorious decades when we were the great heroes because we were on a declining cost-curve, did look to the future. There was a compact to provide a safe, affordable and reliable stream of power, and we looked 10 to 15 years ahead when we made investments. We were probably in a much more speculative business than we realized at the time, but some of the most beneficial investments this country has ever made are in hydroelectric projects that are still there and still turning out very, very low cost electricity.

I think that it is important that we stop now and ask ourselves, "Competition for what?" After all, we still call ourselves public service companies (most of us), and we have public service commissions. What is it that the public demands of us? What is it that 15 or 20 years hence, people will say should have been done? Is price competition and short-term marginal price competition really what it is all about in this industry?

The Electric Industry in Transition

For decades this industry took positions exactly the opposite of what its consumers were voting for. Electric utilities opposed every amendment to the Clean Air Act while our constituents were voting 80 to 85% in favor of them. We made fun of the concerns about nuclear safety before Three-Mile Island. We have made a lot of mistakes in terms of our relationship with our customers, and only in the last decade are our hats turning from black to grey.

We are on the verge of advocating the things that society is most interested in: cleaning up the air, displacing imported oil, focusing on the bills that consumers pay rather than the rates, and actually lowering their bills. We are on the verge of becoming popular with our customers and making customer service really meaningful words in our lexicon. Now, all of the sudden, we are supposed to forget all that and focus on short-term, marginal prices and engage in what we used to call "price wars" at the gasoline filling stations.

Of all the resources on this planet, energy provides perhaps the most vivid example that market competition considers more than just price. We cannot decide that the externalities are zero. That would take us back to the era when the Tennessee Valley Authority was refusing to put provisions in strip-mine coal contracts to reclaim the land lest it raise the price of coal and the American Electric Power System was taking out full-page ads attacking scrubbers.

This industry has had a history of being oblivious to this small planet that we live on and the resources that sustain life, but we have overcome that. Over the decades we have learned that power must be produced safely, with minimum pollution of the air and with respect for the fish in the water and the beauty of the landscape. To assure a cleaner and safer energy future, we must invest in perfecting new sources and new uses.

Energy can pollute, yet it is central to our quality of life. It can cause recessions, yet it is vital to our economic well-being. It can cause us to go to wars and it has. It is time that we help wake up this country to the fact that we are importing over 50% of our oil. This industry has a solution. We now are on the cutting edge of providing, I think, the most important innovation in this country's energy history in a long time: electrification of the transportation sector and production of that electricity from domestic sources. It will take some investments, and it will take a whole lot more from top management than just a head-long, herd-like dash toward price competition only.

Chapter 2: An Unguided Missile?

Certainly competition is a vital force that we must embrace. As one that competes day in and day out, I certainly do not oppose it; I endorse it. But it is only a piece of our job, and it needs to be fitted into a comprehensive policy and program which recognizes that the things that count most for our customers are not counted in the price of electricity, not counted in the price of gasoline today, or not counted at all! We must invest in measures to use our energy more efficiently and cleanly.

It seems to me obvious that the first casualty of this mad dash toward price competition is going to be the consumer. Consumers, as we all know, pay bills, they do not pay rates. If we cut out our efficiency programs and we just pay attention to price, the person who is going to be hurt is the consumer who pays the bill. The efficiency investments that will be eliminated are those that benefit the customer. No one can argue that a customer who reduces his or her consumption of electricity by 20 or 25% is not better off even if the rates go up 5 or 10%. If all the customers have an opportunity to participate, and if you really have a comprehensive program, the issue of equity among customers is not an issue.

DSM investments, in most cases, are cost-effective for the utility. No one can argue that a utility that invests in DSM today, and avoids more expensive supply-side investments tomorrow, is not also better off. It is just simple arithmetic.

I think that it is also important to recognize the difference between the wholesale power market and the retail power market. Among the wholesalers, electricity is just a commodity and competition among generators is a very good thing. We already have wholesale competition in New York State to a considerable extent. If we really want to have price competition that is in harmony with the other needs of society and the electric power industry, then we should forget about retail wheeling and start down the road that is much more sensible for a competitive industry, that of divestiture.

We ought to separate the generation, transmission, and distribution functions of the industry. We ought to create a New York electrical thruway in which all the transmission is managed by a single entity and independent companies take over the distribution function.

At the retail level, customers are interested in their bills, they are interested in cleaner air, and they are interested in the future. And the future, as Alfred Kahn has observed, is primarily in decentralized technologies.

Distributors are particularly well suited to market and support decentralized technologies. I think that model is the one we should be looking toward rather than this crazy quilt coming out of California. California's retail wheeling proposal needs to be put aside and allowed to rest in peace.

What is it that we should be aiming for? There is a tremendous amount of intelligence in the industry and among its regulators, working desperately to produce a perfect model for just price competition. I hope we can enlist a good part of that intelligence to put us on track towards a sustainable energy future that is in harmony with this planet on which we and our children and our grandchildren live. To assure a cleaner and safer energy future, we must invest in perfecting new sources and new uses. That, I think, is the burden that we have. If we think of energy efficiency in the grandest sense of the word—including the replacement of petroleum in the transportation sector with electricity, which is a giant step forward in energy efficiency as well as in solving two of the nation's greatest problems: air pollution and national energy security—then I think that we will be marching down the right road.

Competition can help us move in that direction, but only if it is orchestrated in such a way as to allow investments today for a better tomorrow. The utility industry's initiatives in the solar area, led by Mr. Vesey of Niagara Mohawk, have contributed to developing a market for photovoltaic cells large enough to bring the price down. These market transformation efforts are making a difference, and if we keep the momentum going the supply and demand curves will cross; we will have photovoltaic electricity for peaking purposes in New York State that will be competitive in the next century. The same trend is occurring in fuel cells. The cost of producing a fuel-cell is getting lower and lower. We also have a large biomass resource in New York that should be harnessed.

Our goal should be that energy efficiency will satisfy all the load growth in New York and in the nation. This is absolutely possible. California is meeting all of its growth with energy efficiency. We can learn to become more efficient in the use of energy at the rate of 2 to 4% a year indefinitely, especially if we think in terms of displacing the internal combustion engine system, which has a very, very low efficiency, with electric motors and an electric power system. We can keep the total energy demand in this country as flat in the next two decades as it was between '73 and '85, when it was flatter than a pancake and the economy grew 30%. If you think of a future where total energy demand is flat and we steadily replace the fossil fuels with renewable sources and move toward

Chapter 2: An Unguided Missile?

a solar/hydrogen economy, which will be more and more electric as the trend has been, it is not an impossible dream. But it is certainly a challenge of enormous proportions.

It is a tremendous marketing challenge for the electric power industry to develop more cost-effective technologies that are environmentally satisfactory and that meet the needs of society. That is what we ought to be debating.

The goal of competition should be to produce an industry that is more responsive to society's demands. We have laws in the books that will keep us from going back to our attitudes of the '50s. But we have a chance to rise above the bare letters of the law. The global warming issue is an issue to which this industry has a solution. The issue of imported oil, on which this country has absolutely fallen asleep, is the issue that we ought to be raising. We ought to be spending most of our time talking about real issues such as, "How do we market electric vehicles, electric buses, and electric transportation?" rather than the arcane subjects on which some of the retail competition advocates have been focusing.

I will conclude by quoting someone rarely referred to by a utility executive, but I think the quote is appropriate to this industry's current situation, its responsibility to its customers and to this planet, and to our tremendous challenge. I want to quote from a former New York senator, Robert Kennedy, in a speech that he made on May 5, 1967:

> Let us be clear at the outset that we will find neither national purpose nor personal satisfaction in a mere continuation of economic progress and an endless amassing of worldly goods. We cannot measure national spirit by the Dow Jones Average, nor national achievement by the gross national product.
>
> For the gross national product includes air pollution, and advertising for cigarettes, and ambulances to clean our highways of carnage. It counts special locks on our doors and jails for the people who break them. It includes...the broadcasting of television violence to sell goods to our children.
>
> And if the gross national product includes all of this, there is much that it does not comprehend. It does not allow for the health of our families, the quality of their education, or the joy of their play....It does not include the beauty of our poetry, or

the strength of our marriages, the intelligence of our public debate, or the integrity of our public officials....The gross national product measures neither our wit nor our courage, neither our wisdom nor our learning, neither our compassion nor our devotion to country. It measures everything, in short, except that which makes life worth while; and it can tell us everything about America—except whether we are proud to be Americans.

We must not be fooled by narrow economics. Price competition is only part of the answer. Advancing renewable resources, energy efficiency, and new technologies such as electric vehicles is just as important, perhaps more important to society, than the price of energy. If the price, or some surrogate for price, does not reflect these broader societal concerns, then lower prices today are a false bargain tomorrow. This industry has always looked out for tomorrow.

CHAPTER 3

Competition in the Electric Industry Is Inevitable and Desirable

Alfred E. Kahn
Special Consultant
National Economic Research Associates, Inc.

My answers to the two questions of whether competition in the electric industry is inevitable or desirable are based on two elementary principles—one of economics, the other of physics.

◆ The first is that sunk costs have no relevance to rational decision-making.

◆ The second is my economic version of the physical law that nature abhors a vacuum: society abhors big gaps between prices and marginal costs.

The first principle requires qualification as it applies to the perspective of the people most concerned with the current state of the electric industry: sunk costs are irrelevant except to utility companies and their regulators.

The second principle, similarly, requires qualification as it applies to the real world: Just as a feather and a brick will fall equally fast in a vacuum (or so they say; I don't really believe it) but not in the real world, so while society ought to abhor equally gaps both positive and negative between marginal costs and price, what it actually abhors especially are prices far above marginal cost. In general, only economists and environmentalists experience acute distress when prices are below marginal cost.

It is the second, physical principle, that summarizes my answer to the first question—"is competition in the electric industry inevitable?" The answer is "yes, probably." But for a full answer, we have to begin at the beginning.

The changes that are going on in the electric industry are part of something much broader that is going on all over the world, from the United States to Tanzania, from Argentina, Mexico, Peru, Colombia and Chile to Spain,

Portugal, the UK, Eastern Europe, New Zealand and Australia, with China and even India in between. It is, to put it very broadly but I think profoundly, the breaking up of centralized planning, cartelization and protectionism and the triumph of the market. The degree and extent of the change have varied enormously from one country to another, from the revolutionary transformations in Eastern Europe and privatizations in both hemispheres to partial or thoroughgoing deregulations of individual industries in the United States—trucking, the railroads and airlines, telecommunications, oil, gas, financial markets and now the electric and gas utilities.

The motivating factors have, however, varied widely from one industry or market context to another. In telecommunications, above all else it has been the revolutions marked by the commercialization of microwave, fiber optics, the convergence of telecommunications and computer technologies and the geometrically declining costs of switching: it was simply impossible to continue to confine the exploitation of that kind of burgeoning technology, with its exponentially expanding diversity of possible services, to a single monopoly no matter how efficient, no matter how many Ph.D.s on its payroll.

If in telecommunications the root cause has been technological explosion and progress, in some sense the explanation in electric power has been technological and institutional failure. I cannot believe that we would be witnessing the disintegration of the tightly-regulated franchised monopoly system in electricity if the industry had continued to perform in the '70s and '80s as it did in the '50s and '60s, when the average retail price in the United States declined more than 40 percent in real terms.

Clearly, the trauma that produced the change was in many ways unique to the electric industry; it was a combination of inflation, with its particularly heavy effect on the costs of highly capital-intensive industries; the explosion of fuel prices after 25 or 30 years in which they had gone down continuously in real terms; the nuclear fiasco; the sudden exhaustion—indeed, reversal—of what we had thought was an inexorable increase year by year in the economies of scale, for a large number of reasons, institutional as well as technological; and the abrupt slowing down in the growth of demand, with the result that companies, having committed themselves to building large, long-lead-time generating plants, suddenly found themselves with large amounts of excess capacity, constructed at grossly inflated costs.

Chapter 3: Competition in the Electric Industry Is Inevitable

And so the industry in the 1980s confronted the very situation that my political-economic version of nature abhorring a vacuum contemplates—high-cost plant entering into rate base, excess capacity and, at least equally important, a collapse of the prices of oil and gas all pushed marginal costs, both short and long-term, below the utility companies' average total costs or revenue requirements. I am not the one to present you with definitive estimates of the size of this gap, but our Niagara Mohawk hosts have, in a recent PSC proceeding, submitted figures showing their average residential price in 1992 of 10.58 cents a kilowatt hour and industrial prices of 7.9 cents, excluding its allocation from the New York Power Authority, and 5.5 cents including that allocation, as compared with levelized full costs of new combined cycle plants in the 5 to 6 cent range.[1] The last figure accords with estimates of long run incremental cost by my colleagues Miles Bidwell and Lewis Perl, but Perl also cites estimates of short-run marginal generating costs of only 2 to 3 cents a kilowatt hour and says that many IPPs "could produce electricity and make a profit at 4.0 cents a kilowatt hour."[2]

These huge gaps between true economic costs and regulated rates have given rise to an irresistible temptation for big buyers to shop around and for generating companies to seek business out-of-territory. So we have a large, increasing number of transactions between willing sellers and willing buyers crossing previously inviolable geographical market boundaries; and, as a consequence, incumbent utility companies—their previously exclusive franchises subject to invasion—are desperately worried about the prospective stranding of hundreds of billions of dollars of investments and their ability to recover the associated sunk costs, costs that first economic principles tell us should be ignored.

These fears are compounded by the huge disallowances by regulators during the 1980s of costs sunk in expensive base load plants—which have also had a permanent effect on the willingness of utility companies to undertake long-lead-time investments. We used to worry about the Averch-Johnson distortion under rate-of-return regulation—the

---

[1] "The Impacts of Emerging Competition in the Electric Utility Industry," April 7, 1994, submitted to the New York Public Service Commission in its Case Nos. 93-E-0376 and 0378, pp. 27 and 31.

[2] "Rewriting the Rules of the Road: Retail Wheeling and Competition in Electric Generation," April 25–26, 1994, p. 2.

incentive to build costly plants in order to have a bigger rate base, on which companies would be permitted a return. In the '80s, the industry experienced, in effect, a reverse Averch-Johnson incentive—a fear of expanding rate base and particularly of risky long-lead-time investments, which I think played an important role in making many electric companies complaisant about other people undertaking responsibility for investing in new generation.

Whatever the importance of this reluctance on the part of the utility companies, they have in any event been compelled, by the Public Utility Regulatory Policies Act of 1978 (PURPA) and state regulators enthusiastically administering that Act and increasingly requiring competitive procurement, to purchase increasing proportions of their power needs from independent suppliers. This has produced the dramatic result that the share of new national capacity contributed by non-utility generators increased from 14.8 percent in 1986 to an average of 49.8 percent in 1988 through 1991 and 88.4 percent in 1992.[3] These rapidly accumulating purchase obligations, often at artificially high prices dictated by state regulators and legislators or on the basis of estimates of avoidable costs that have turned out to be much too high, have evidently made a multi-billion dollar contribution to increasing the yawning gap between the utility companies' rates and their marginal costs. And that gap, in turn—as I have already pointed out—is what has led to the ballooning out-of-territory transactions, at bargain rates, that threaten to leave stranded those very, inflated investment costs that have themselves produced the gap.

If, however, the explanation of the radical institutional changes in the electric industry is rooted largely not in technology but in adventitious, transitory circumstances of the '70s and '80s, if it is in some way the consequence of abrupt changes in the relationship between marginal costs and average total costs, is it possible that the trend may reverse itself?

I think that is unlikely, for several reasons. One, of course, is the virtual universality of these historical trends. That is why I began by pointing them out. Second, it is my impression that the trend of generating technology is not propitious to a return of monopoly. Instead, it appears that the major developments on the horizon—the combined cycle gas turbine,

---

[3] Edison Electric Institute, "Capacity and Generation of Non-Utility Sources of Energy," 1989 and 1992.

small scale fluidized bed, fuel cells and co-generation—are largely centrifugal rather than centripetal.

Third, for all the differences in local, industry-specific circumstances, regulation and deregulation have one characteristic in common: once begun they have an almost irresistible tendency to become more and more pervasive, to the point of universality. This is a phenomenon that I have documented at length in other contexts. Effective regulation of airline prices, for example, necessitated regulation also of the commissions the carriers might pay to travel agents, their charges for in-flight entertainment, the frequency of their flights and even (in the international arena) the size of the sandwiches they served. Similarly, deregulation of entry into individual markets soon required a corresponding lifting of the barriers in all markets, relieving incumbents of the obligation to serve, deregulation of price and of all the other inducements to attract travelers: head set charges, travel agents' commissions, frequent flyer credits, free upgrades to first class, and so on. The unravelling of telecommunications regulation over the last 35 years has been a similar, cumulative process, which will almost certainly continue over the next decade.

One of the main reasons the process tends to be cumulative and self-reinforcing is that once you begin to loosen the bonds and admit competition selectively, it introduces strains and distortions that can typically be resolved only by further deregulation. That was why extending to individual airlines the right to enter individual markets soon led to extending it to all carriers in all markets; conferring that freedom on a discretionary basis (competitors had to be free to enter but not compelled to do so) inevitably required relieving incumbents of the obligation to remain: freedom of entry necessarily implies freedom of exit. Similarly, removal of restraints on entry would quickly have produced a gross inflation of inefficient non-price competition had it not promptly been accompanied by removal of restrictions on price competition as well; and that in turn necessitated deregulation of all the other ways in which carriers might compete and of all the other conditions of service (for example, the number of seats crowded into a plane and the space between them) except safety.

In electric power, the competition that has been admitted has been in large measure distorted by the past and continuing regulation of the incumbent companies. Competitive decisions are being made on the basis of a comparison between true competitive prices and real economic costs on the one side and, on the other, regulated rates that may diverge widely from the marginal costs of the utility suppliers.

A business decides whether to co-generate or not, for example, on the basis of the actual prospective incremental costs of obtaining its power in that way. Similarly, when a big buyer shops for supplies out-of-region, it is looking at prices set by competitive forces at something approximating marginal costs. But, in making those decisions, the departing buyers compare those economic costs with regulated rates that are set to recover the local utility company's average revenue requirements, which have a very large component of sunk, historic costs. And, as I have just pointed out, that overlay of inadequately depreciated historic costs, in the presence of excess capacity, produced average revenue requirements in the '80s and early '90s far above the true marginal costs of the utility companies themselves. In these circumstances there was and still is no way of knowing to what extent the competition was efficient, to what extent inefficient.

A further distortion has been introduced by the incumbent companies' obligation to serve, as providers of last resort to all customers, and to incur the cost, largely sunk, of fulfilling that responsibility. The introduction of competition offers buyers the opportunity to escape being charged for those sunk costs by taking advantage of competitively priced power available elsewhere, while, however, retaining the option of returning eventually and taking advantage of their local utility's continuing obligation to serve. So their decisions to pick up low-cost supplies elsewhere or to co-generate may be economic only on the assumption that if the co-generating facility fails or the cheap supplies elsewhere dry up, the buyer may be able to return and demand service, without penalty. So, again, differential obligations and consequent differences in regulatorily-imposed cost burdens may produce distortions in competition among the several rival suppliers.

In these various ways, the partial introduction of competition into a thoroughly-regulated utility industry creates distortions and strains, which can typically be resolved or eliminated only by extending the freedoms symmetrically to all competitors. Specifically, this means giving the local utility companies freedom to price competitively and freedom from the continuing obligation to provide firm backup service to customers who exercise their freedom to escape the costs of providing that option. And that, I am convinced, is the historical process in which we find ourselves inevitably engaged—inevitably, that is to say, unless and until we encounter another unforeseeable historic conjuncture, either specific to this industry or in the economy at large, such as occurred in the 1970s or the 1930s.

That is why my answer to the first question, is competition inevitable?, is "Yes, probably."

My description of the distortions created by partial, asymmetrical deregulation has already taken us deeply into my answer to the second question—is it desirable? The answer is "Yes, probably, but...."

I do not have much to contribute to amplifying the first two-thirds of that answer. Experience in industry generally, and in the deregulated industries in particular, amply supports the proposition that, wherever it is remotely feasible, competition is superior to regulated monopoly as an institutional mechanism for producing close attention by managements to efficiency and promoting progress, both in methods of production and in offering consumers an ever-expanding variety of choices, at efficient prices. And it does indeed appear that competitive generation of power and the prospect of its benefits becoming available to more and more customers has already subjected electric utility companies to salutary pressures to cut their costs; to figure out what they can do well and to do it, and to stop doing things that they can't do well—for some this means leaving generation to outsiders, for others, going into generation all over the country; to offer an increasing variety of bundles of services, such as has occurred in telecommunications and airlines—interruptibles are one such bundle or set of bundles; to become increasingly assemblers of services or system integrators rather than supplying all the services themselves: that really is the significance of acquiring energy supplies by competitive bidding.

I have no way of weighing those benefits of competition against the benefits in principle of centralized responsibility for reliability of supply, coordination of investments and operations and wholehearted cooperation, such as occurs in power pools among non-competing, geographically separated franchised monopolies. The catastrophic errors of the '70s under the historic system, however, the apparent persistence of inefficiencies in regulatorily-prescribed rate structures (which competition tends to erode) and the apparent evidence from Great Britain that a non-vertically-integrated system can work all help produce my "Yes, probably" answer.

The Electric Industry in Transition

The "but" that I have tacked on to that response is important, however. It has a number of aspects.

The most important one is the necessity of adapting regulation so as to ensure that the competition is efficient. Competition between unregulated options, at prices equal to marginal costs, and regulated rates that diverge substantially from marginal cost provides absolutely no assurance that production is being distributed among the rivals on the basis of their true economic costs and, therefore, is at maximum efficiency. How then do we eliminate the distortion?

The inflation of the costs and rates of the utility companies that is, in turn, the source of the distorted competition has two kinds of sources. One is the continuing, current burdens imposed on them, but not their competitors, to carry the burdens of continuing to serve as suppliers of last resort; of financing regulatorily-prescribed cross-subsidizations; of promoting and subsidizing conservation on the part of their customers; and of recognizing environmental costs from which their unregulated competitors are free. These distortions are unequivocal and continuing, and the only acceptable solution is to eliminate them by:

- permitting the utility to sell back-up or option-demand services separately or—what comes to the same thing—require customers that want the option of shifting to competitors to pay separately for the right to return and demand service;

- eliminating all artificial, regulatorily-imposed requirements to continue to buy independently-generated power at any rates other than avoidable costs, estimated as honestly as possible, or to subsidize energy conservation (as long as rates do not fall short of true marginal costs);

- rebalancing rates to eliminate cross-subsidizations or taxing all suppliers proportionately to carry those costs; and

- eliminating all environmental cost burdens that bear only on the utility companies, or, preferably, extending them neutrally to all supply sources, regulated and non-regulated.

The other—and I suspect by far the larger—source of the discrepancy between utility company rates and marginal costs are the sunk costs. These, I am informed, are running in the hundreds of billions of dollars, and are predominantly comprised of the heavy costs of the generating plants inherited from the '70s and early '80s and the contractual commitments to purchase non-utility-generated (NUG) power at prices in excess of the utilities' own present short-run marginal costs.

Beyond pointing out that economic efficiency would best be served by ignoring these sunk costs and freeing the utility companies to reduce their rates to marginal costs, to the extent necessary to meet competition, I have no particular enlightenment to offer on the question of the companies' entitlement to recover those, presumably prudently-incurred costs—a large portion incurred, indeed, on orders by the regulators—or about the equity or efficiency of permitting some of them to be shifted from escaping to captive customers, except to the extent that the latter have heretofore been the beneficiaries of cross-subsidization at the expense of the former. But I do have several unequivocal observations to urge upon you with respect to the method of their recovery.

First, to the extent that the incumbent electric companies are able to recover from competitors the same proportionate contribution as was and remains embedded in their own rates for the business that they lose to those competitors, by marking up their charges for access to bottleneck transmission and distribution facilities, there is absolutely no conflict between that recovery and economically efficient competition: all that it requires, for competitors to survive or fail depending solely on whether their marginal costs are equal to or lower than those of the incumbents, is that the utility companies incorporate that same markup over marginal costs in their own retail prices as well. Indeed, the exaction of such a equivalent contribution is not only not inconsistent with efficient competition, it is *essential* to ensuring that the competition is efficient while also permitting continued recovery of the stranded costs. The frequent observation by would-be competitors and often, unfortunately, by regulators themselves that such levies "discourage competition" is at best extremely misleading. Setting aside what we might call "infant industry" considerations (to the effect that it may contribute to achievement of the long-term benefits of competition to confer special preferences or advantages on new competitors or correspondingly special handicaps on the incumbents), an equi-proportionate charge or mark-up above marginal costs for competitors and incumbent utilities alike is fully compatible with efficient competition between them.

The compatibility of such a levy on competitors and efficient competition is not dependent, in principle, upon whether the competitors are or are not dependent upon the utility company for access to an essential facility. The latter circumstance merely determines whether it is possible to equalize competition between them in this way, while also recovering stranded costs. An electric company is hardly in a position to impose such levies on customers who are induced by the excess of rates over marginal costs to leave the state or to install insulation that they would not otherwise install. But the principle is the same: to the extent those latter decisions are induced by the excess of price over marginal costs, they are inefficient. If the alternatives could be taxed proportionately, it would promote economic efficiency rather than the reverse. In the same way, regulatory commissions are considering alternative methods of taxing all rival providers of telecommunications services proportionally, in order to finance the continuing cross-subsidization required to maintain universality of subscription, as the ability of telephone companies to recover such costs in their charges to competitors for access to their local networks is increasingly limited by the ability of those competitors and their customers to bypass those facilities. Meanwhile the best pragmatic prescription, for both the telephone and electric companies, would be to get as much of those costs back as possible, writing down the book value of inadequately depreciated plants, while the getting is good.

But the recovery of sunk costs in these ways, it is worth reemphasizing, conflicts with first-best economic efficiency because it holds prices above marginal costs. While therefore it remains fully consistent with the competition being efficient, it prevents competition from achieving its other purpose, of driving prices down to marginal costs.

While the political consequences of divergences between prices and marginal costs may be asymmetrical, depending upon the algebraic sign of the difference, the economics is not. I have in the past criticized the tendency of regulatory commissions to play the game of heads-we-win, tails-you-lose—disallowing large portions of the costs of unsuccessful investments in nuclear plants on the ground that the facilities were either not "used and useful" or that their true economic value was lower than book, while continuing to value successful plants at depreciated original cost. John Rowe, President of New England Electric System, has pointed out that the same accusation may properly be levied against competitors and customers who enter into transactions for wholesale power at market value, when their utility rates exceed marginal costs, while demanding the right to use the utility companies' transmission and distribution facilities at rates regulated on the basis of book costs. Both equity

Chapter 3: Competition in the Electric Industry Is Inevitable

and first-best economic efficiency require that, just as the price of the power itself be driven down to marginal cost, the rates for its carriage be raised to that same target, to the extent they are at present below.

I have not the slightest idea to what extent the latter correction would offset the former, in terms of the ability of the utility companies to continue to recover their stranded, sunk costs. But to the extent there remains a residual problem (and I suspect it is or would be a very large one), I am going to refrain from offering to the utilities the cheap academic advice that to an economist sunk costs are bygones and best forgotten. That is too easy for me to say: those are your sunk costs, not mine!

Part 2

The Electric Industry in Transition

INTRODUCTION

What Does the Competitive Market Look Like and How Will the Industry Work?

William LeBlanc
Director, Strategic Marketing
Barakat & Chamberlin, Inc.

Deregulation is the experiment of letting the market go where it leads itself. Those accustomed to regulation are comfortable with studied predictions of the future, guaranteed markets and profits, and a knowledge of their allies and foes. The competitive market knows very little comfort. Every day a new entrant can arise, a large customer may come or go, a product failure can ruin a quarter financially, or a new service can provide outstanding profits. Every move is a calculated risk, far above those risks seen in the regulated environment.

Economic theorists believe that the free market will result in the greatest economic efficiency if true competition exists. Environmental stewards believe that external costs of energy use will not all be included in the energy price under deregulation. Utility executives ponder ways to maintain shareholder value and market share under the new competitive regime. A certain answer to the question posed for this session's panelists is, "We can only speculate, as the market is more powerful than we are."

The session heard views from three different vantage points: a representative of the already deregulated United Kingdom (U.K.) market, a U.S. generation developer, and an academic economist.

Graham Hadley, managing director for National Power PLC, one of Great Britain's large power brokers, speaks from experience when he outlines the benefits and costs of introducing competition into a government-run utility. The U.K., where all large customers have the ability to choose suppliers, has probably gone further and faster than any other electricity market in the world. One observation that deserves further notice is Mr. Hadley's comment that during the short six-month period during which the old nationalized system turned into the competitive system, a tremendous amount of work was successfully completed. This implies that the industry made many more decisions, developed more contracts, took more risks, and created more business opportunities than would have been possible under regulation. Human energy shifted from making regulation work to making business work. In short, everybody worked faster and harder because they had no other choice.

The focus of the utilities' business probably changed dramatically from simply creating and distributing electricity to meeting customer requirements for quality, price, and service. If large customers are not pleased with the service level, they can choose to leave. If prices are higher than those of other entrants, an entity loses market share.

We can only speculate as to how electricity service will change with new forms of profit motivation and risks. In the U.S., the primary focus is on the price of electricity and how retail competition will force that price down. There is little consensus on how much the price will drop, although in some high-priced areas of the country it could drop dramatically. While some discussion regarding the consequences to the small customer has arisen, no one knows whether rural, small, or low income customers will be appropriately served under full retail wheeling.

Certainly the telecommunications companies found it worth their time to vie for the residential customer market much sooner than consultants had speculated after the Bell breakup. Airline customers are not always served from smaller towns the way they were before deregulation, but service on the whole has increased tremendously. Small, low-cost entrants constantly put pressure on the larger airlines to cut prices. Also, airlines know that they compete with the automobile for short flights and vacation travel.

Hadley concludes that privatization in England has worked better than expected, with prices dropping significantly except for industrial customers who had previously been subsidized. Whether this will continue when the smaller customers are included in the equation is still a guess. Very little demand-side management is being conducted, however, with the exception of load management-type programs. Mr. Hadley is obviously enthusiastic about the benefits of wholesale competition, but less sure about the additional benefits of full retail wheeling.

Jeanine Hull, from LG&E Power, emphasizes that the industry is working under a dichotomy at this time. To make money today a developer or utility must work under today's regulations and rules, but in order to be appropriately positioned for tomorrow, these same organizations must prepare for a deregulated market with new rules and regulations. Especially for utilities, this puts management between a rock and a hard place. Regulators often keep their thoughts close to their chest until the last minute when decisions are made. The market, somewhat unguided, is moving and regulators will soon find that their power to influence it is largely lost.

Hull notes that regulators must work with close attention to changes in state and federal law. These changes may happen very slowly, but are ultimately necessary. This adds to the uncertainty of the business climate. The movement is clear, however, in the generating industry. Power will be garnered competitively, there will be more suppliers than demand, and the lean, efficient suppliers will be the survivors.

William Hogan, from Harvard University, speaks in great detail about possible and preferred forms of the electric industry under increased competition. He particularly emphasizes the structure of the transmission systems, power pools, and pricing. The area that I think creates the greatest intrigue regards market pricing of power. How will brokers and generators sell their power? Will it be split evenly between spot market and long-term contracts? How long will customer contracts last, and what "out" clauses will customers demand? Will customers like highly variable prices, or will most opt to pay for added price stability? Will residential customers really have choice, or will they be left paying higher prices than they currently do?

Most economists concur that average cost pricing creates potential for inefficiencies in demand. Utilities are obviously nervous about the potential shortfall between current prices, which leads to cost recovery, and marginal prices. Many utilities prefer a slow approach to deregulation in order to maintain revenues to pay for current investments. However, what are the implications for a "go slow" approach? One that is particularly worrisome is that smaller customers will be left holding the bill for overcapacity. I believe that residential customers will have enough political fire-power to receive retail access very soon after large customers. A big question, however, is whether the bulk of residential and small commercial customers have the interest, time, and wherewith-all to go electricity shopping or to make appropriate decisions regarding reliability and other service levels. If they are forced into making these market decisions, what are the transactional costs, and do they approach the magnitude of the benefits of retail access?

As a final reflection on the question posed to this panel, I would like to add some points about the retail energy service market and customer value. In order for customers to truly reap the benefits of competition, it should be kept in mind that electricity is not the only product to think about. The energy service market (delivering heat, light, entertainment, etc.), which is essentially the combination of energy-using equipment and the fuel that drives it, is of ultimate concern. DSM programs are designed to provide the highest energy service level at the lowest cost by balancing the cost of electricity versus the technology cost. While DSM has seen successes in recent years, utilities will be reluctant to pursue DSM as a generation resource option until the regulatory picture becomes clear. Integrated resource planning and an open market for power are largely incompatible.

However, market barriers to the use of energy efficient products still do exist. Many of these barriers are an outgrowth of the monopolistic, vertically-integrated utility structure we have had for so many decades. A viable energy service industry outside of utilities does not yet exist. The DSM role could be played through a state-wide organization that collects money from users of the grid and then allocates the money to cost-effective programs throughout the state. The funds could be distributed on the basis of market barriers costs, and energy service companies and distribution companies could bid for the money. In this way, the competitive market would be involved on the demand side, energy service providers would have incentive to stretch the money over as many customers as possible through customer dollar contributions, and the industry would have time to evolve. Ultimately, as market transformation occurred, the money for those market areas would be withdrawn. A method like this would allow the U.S. to simultaneously reap the benefits of lower electricity prices and improved end-use efficiency.

CHAPTER 4

A Competitive Electricity Market: Lessons from the U.K.?

Graham H. Hadley
Executive Director and Managing Director
of International Business Development
National Power PLC

Introduction

Recent years have seen something like a consensus develop across the world that of all the ways of organizing economic activity, markets represent the most effective means of increasing efficiency and bringing benefits to customers. The electricity sector has been part of this general consensus. In different sorts of countries across the world, electricity markets are being opened up to competition. Often the motive is to achieve the gains in operational efficiency and investment efficiency that are the hallmarks of markets everywhere. Sometimes the motive is to secure access to funds for new investment that cannot be found within a centrally-planned and publicly-owned system. Sometimes, there are wider political goals.

However, whatever the prevailing views about markets as such, moves to liberalize electricity systems have often met resistance from the electricity industries concerned and quite often from governments too. It has been argued that electricity is "different"; that competition is wasteful; that electricity is "strategic," and central planning needs to be retained; that competition would raise costs and prices, endanger security of supply, or prevent wider social or industrial goals from being achieved.

In the light of these concerns, the U.K. experience in introducing competitive electricity markets provides a useful case study. This is not because the U.K. system can or should be simply transplanted to other countries; for example changes in the U.S. electricity industry will have a different starting point, and will reflect the U.S.'s particular commercial ethos and legislative framework. Nor is it because the U.K. was first in the field: competition in electricity generation on a significant scale started in the U.S. some years earlier.

Rather, the U.K. experience is relevant because liberalization has been taken further in the U.K. than in other countries over quite a short time period. In formulating the principles of the new U.K. system, all the generic issues that are argued over in the U.S. and elsewhere had to be faced and resolved. In implementing the new system the conclusions reached and the principles adopted have been tested out, at least over a 3-4 year time period. I believe, therefore, that it is well worthwhile for those involved in the change process in the U.S. to consider what has been done in the UK.

This paper concentrates on the privatization of the electricity industry in England and Wales. The Scottish electricity system, which is essentially self-contained, was privatized at the same time but in a less radical fashion. The physically separate system in Northern Ireland was privatized later and on a different basis, reflecting the much smaller size of that system and the more limited opportunities felt to be available for introducing competition.

The U.K. Electricity Privatization Process

In the 1980s, the British Government embarked on a program of privatizing the major nationalized utilities. Typically, these industries had been brought into state ownership in the years immediately after the Second World War and operated as national monopolies. Although the desire for the financial proceeds from privatization, in order to reduce the public sector borrowing requirement, was an important initial motivation of the Government, subsequently the promotion of efficiency through competition and the promotion of a wider share of ownership became more relevant. The first utility privatizations, of the telecommunications and gas industries in 1984 and 1986, essentially moved monopoly organizations from the public sector to the private sector, whilst providing a framework within which competition could develop. However, by the time the Government announced in 1987 its intention to privatize the electricity industry, the introduction of competition from the beginning had become an important objective.

The broad time scale of electricity privatization is shown in Chart 1. The privatization process followed a number of key phases.

The Debate on Industrial Structure

The objectives of Government and the views of the industry—or at least of the Central Electricity Generating Board (CEGB), which was responsible for generation and transmission in England and Wales—were far

Chapter 4: A Competitive Electricity Market

Table 6-1. Electricity Privatization in England and Wales

1987/88	1988/89	1989/90	1990/91
What sort of industry?	What sort of market?	Restructuring the industry	Flotation
	What about Nuclear?	The market rules	

from being compatible. The Government wanted to introduce competition immediately and felt that the CEGB needed to be broken up to achieve that. The CEGB argued that the separation of generation and transmission would result in significant extra costs and would jeopardize security of supply.

Both Government and the industry looked around the world at possible models for the U.K. There were various examples of fragmented systems; but these operated on the basis of cooperation rather than competition. Such examples of competition as there were—and the U.S. was the most important example—were mostly of competition grafted onto a monopoly utility base.

Having studied overseas models, planners found it necessary to determine the general form for the privatization in the U.K. This was announced in 1988 and, as discussed below, involved the break-up of the CEGB and the opening of the market to new entrants.

The Nature of the Market
Once the decision was made that competition should be introduced into generation and supply—while still preserving the benefits of despatching plant according to a merit order of some kind—it was necessary to determine which form of market would best achieve these ends and to decide the general nature of the initial commercial arrangements. Since there was no obvious existing model to adopt, a major effort was put into the analysis of options, debate and negotiation. In the event, a decision was made in the autumn of 1989 on the pooling arrangements that would be adopted and the fundamentals of the initial contracts.

The Nuclear Issue
The Government originally intended not only to include nuclear power stations within the privatization, but also to ensure that the development

of nuclear power continued—specifically that the existing CEGB program for four pressurized water reactor (PWR) stations be carried forward.

The nuclear issue affected the form of the privatization: it required that a company be formed that would be large enough to provide a credible technical base for nuclear operations and be able to manage the financial strains of nuclear development. It also resulted in arrangements being put in place to oblige supply companies to contract for a period ahead for a quota of nuclear capacity.

Considerable effort was devoted over a period of 18 months to the search for a means of achieving the development and operation of nuclear power within a competitive market. In the event, it was clear that potential investors would not accept the financial risks involved without Government underwriting of these risks to an extent that the Government, in its turn, was not prepared to accept. During 1989 first the older generation of nuclear stations and then the remaining stations were removed from the privatization, to be placed in the hands of companies which would remain (at least for some years ahead) in state ownership, with the protection of a subsidy from electricity consumers.

Creating the New Market

In the autumn of 1989, when the market structure was determined, "Vesting Day"—when the assets and liabilities of the old nationalized organizations were to be vested in the new companies and the new market was to start operating—had already been fixed for 31 March 1990. That gave the industry six months in which to finalize the codes and agreements that constituted the rules of the new market, agree on the initial contracts for electricity and fuel purchases, determine the final allocation of assets, and establish the new companies as going concerns.

The amount of work that was undertaken during this period was immense; it is interesting to reflect on what can be achieved in a short time when the will to achieve it is there.

Flotation

There were certainly some who doubted whether a change of such magnitude over such a short time scale would work. However, the lights stayed on; and although some detailed issues had been left to be resolved after "vesting" it was clear that the new industry was working well, and that the structure was sound. It was possible for the Government to move quickly to float the new companies on the Stock

Chapter 4: A Competitive Electricity Market

Exchange—the first flotation, of the regional electricity supply companies, taking place late in 1990.

The Restructuring of the Electricity Industry

On the basis that competition was to function in both generation and supply, some key principles were settled at an early stage. In particular, there would need to be vertical disaggregation to separate the functions of generation, transmission, distribution and supply. The distinction between distribution and supply—in other words, between the monopoly activity of operating a local network and the potentially competitive activity of buying electricity and selling it to final consumers—was not fully realized at the outset, but it became readily apparent that separation was necessary if effective competition in supply was to develop. As with generation and transmission, it was recognized that it was important to prevent ownership of the monopoly activity from distorting or preventing competition.

It was acknowledged that the monopoly activities of transmission and distribution would need to be regulated indefinitely, and a regulatory system was established on lines previously adopted for the telecommunications, gas and water industries, with prices capped under an "RPI-X" (inflation minus an efficiency factor) formula—there is no control on profits as such. Generation was not subjected to price control. Supply was brought under some controls, though this was seen as a transitional measure.

Various measures were put in place to ease the transition to the new competitive market. The most important of these were:

♦ A phased roll-out of competition in supply. Initially around 30% of the market was opened to competition for customers with a maximum demand of 1 MW or more. This was increased to 50% of the market in 1994 (customers with a demand in excess of 100 kW). The whole market will be open to competition in 1998.

♦ Initial three-year contracts for the purchase of electricity by the regional electricity companies (RECs) for a substantial proportion of their market (technically these contracts were hedging instruments) and for the purchase of coal by the main fossil generators.

♦ One-year contracts at a special rate for a small number of very large industrial customers, as a temporary continuation of a previous scheme.

The Electric Industry in Transition

- Nuclear support arrangements for an eight year period: an obligation on the RECs to contract for a certain amount of nuclear capacity and a corresponding levy on all suppliers to compensate the RECs for the higher purchase price of nuclear-generated electricity.

In organizational terms, major changes had to be made to the monolithic structure of the electricity industry in England and Wales. The CEGB had an effective monopoly of generation and transmission (although independent generation was permitted, the commercial framework made it unattractive). Distribution and supply were the responsibility of twelve regional area boards, and an electricity council existed as a consultative and coordinating body. The industry was state-owned, and the Government exercised some key powers in terms of setting financial targets, approving major capital investments and appointing Board members.

The restructuring is shown in Chart 2. The CEGB was split into three generating companies (two for fossil fuels and one for nuclear energy) and a transmission company (the National Grid Company). The task of establishing these as self-contained companies was a major one, though

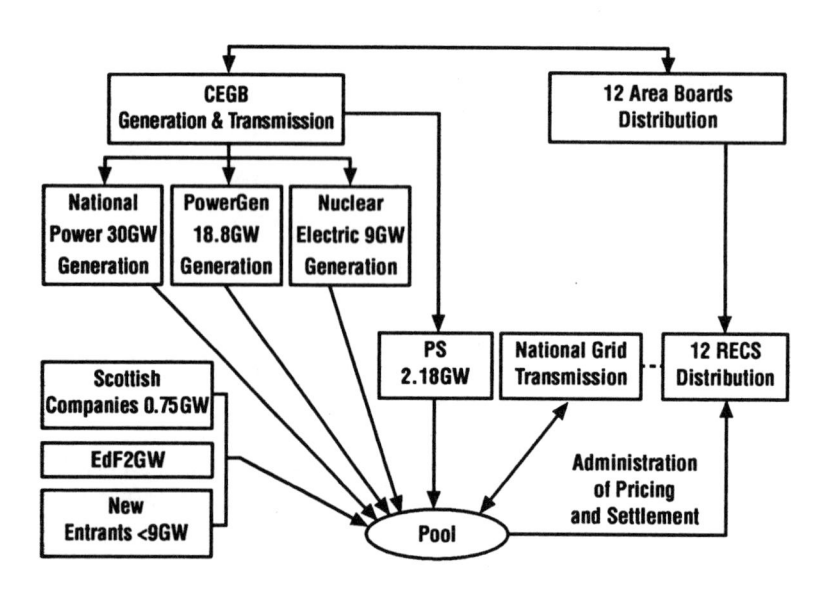

Figure 6-2. Privatization of the Electricity Supply Industry in England and Wales

achieved remarkably quickly. The area boards became regional electricity companies. New entrants were free to enter both generation and supply. A regulatory agency ("OFFER") was established.

The new market itself has a number of components.

- ◆ The central scheduling and despatch of plant, primarily on the basis of prices bid in for each generating set.

- ◆ An electricity pool, into which generators sell their output and from which suppliers buy electricity to sell on to final customers.

- ◆ Pool prices set each half hour by the price of the marginal plant dispatched (plus a charge for system costs and, when relevant, a capacity element) and a settlement system to administer the cash flows involved in pool operations.

- ◆ A set of codes and agreements that together constitute the market rules, set the technical requirements for connections to the system, and determine the basis for commercial transactions.

The Results of Liberalization and Privatization

The results of restructuring and privatizing the electricity industry in England and Wales have been remarkable by any standard. To take the effects on National Power, for example:

- ◆ The pressures of competition have forced us to fundamentally review all our main activities—what we do and how we do it—in order to drive costs out of the business, become a least-cost producer, and operate in accordance with the world's best practice.

- ◆ The results in efficiency and cost terms have been startling: over four years turnover per employee has virtually doubled, productivity has improved by 84% and staff numbers have declined from over 17,000 to around 6,000. These savings have enabled the Company to increase profits and dividends even while its market share fell by 13 percentage points, while competitors entered the market, and while absolute sales also declined.

- ◆ The ethos of the company has changed from that of a public-sector utility to that of a commercially-orientated energy company.

♦ The company has internationalized: it now sees itself as a major player in the global independent power-generation market and is developing projects in a range of countries around the world, including the U.S.

Electricity customers in England and Wales have benefited from this competitive activity. Since privatization, wholesale prices to the vast majority of large customers have typically fallen by around 16% in real terms and prices to domestic and small consumers by 8%.

Naturally, not everyone is pleased with developments—in particular those very large electricity consumers who have lost the benefit of the subsidies they previously enjoyed, and British coal industry interests who have seen their protected status removed and demand for coal reduced as both the successor generators and new entrants bring in gas-fired plant. In addition, there have inevitably been some distortions in the market caused by the transitional arrangements that were put in place. However, these are temporary effects that will work themselves out over time.

Liberalization in the U.K. and U.S.: A Comparison
The pace of change in the U.K. has been dramatic; in the U.S., though liberalization started earlier, progress has been relatively slow and the traditional utility structure is still intact (if somewhat ragged at the edges).

There are some obvious reasons why change in the U.K. could happen at a faster pace than would be possible in the U.S. The U.K. industry was in single (Government) ownership and could be restructured in a way that would be virtually impossible with a more diversified shareholder ownership. The system in England and Wales is linked to that in Scotland and France but nevertheless is much smaller than that in the U.S., has a geographical coherence, and is strongly interconnected internally. The U.K. is a unitary, not a federal, country; it has single-tier regulation—which reduces the scope for delays.

However, in other ways the climate in the U.S. is more favorable to liberalization and deregulation than that in the U.K. The benefits of electricity markets are clear—for consumers if not for monopolists. There is a prevailing market ethos in the U.S.—more so than in the U.K. or in most other countries. In the U.S., the right of the consumer to choice is strongly defended, and the number of goods and services not subject to choice has been progressively reduced over the years. In the electricity sector, the U.S. has established a successful track record in introducing independent generation alongside a utility system. There is the gas precedent too. Further liberalization in the U.S. is inevitable; but it may well happen as a process of evolution rather than as an act of political will.

Assessment of Alternative Options

Different countries are approaching liberalization of electricity markets in different ways. For simplicity, it is helpful to consider two models, which I will term full liberalization and partial liberalization. The key features of the two models are as follows:

- Full liberalization relies on market forces to the maximum extent. There is no central planning, but competition in both demand and supply, freedom of entry and exit, access to networks on an equal basis for all competitors, and non-discrimination provisions exist.

- In partial liberalization central planning and supply are retained, allied to the transmission system operator, but there is competition for new generation capacity with associated long-term contracts and distinction between utilities and IPPs.

The full liberalization model is the one that has been adopted in England and Wales, though the supply competition element is being phased in.

If one accepts the general case for markets and competition over central planning and monopolies, then the obvious conclusion would seem to be that the full liberalization model is likely to bring the greater benefits to consumers. However, the electricity industry has its specific features and before embarking on a major process of change it is important to consider how any market model would work in practice. We can set down some efficiency criteria against which different competitive models can be judged.

- Investment efficiency: does the system promote the right amount of new investment (and of disinvestment) of the right type, at least cost?

- Is security of supply maintained?

- Is the least cost-generating capacity available called on to generate on any day?

- ◆ Does the system promote maximum managerial efficiency, so as to minimize costs within the industry?

- ◆ Do efficiency increases and cost reductions feed through to minimize prices to consumers? Is the structure of prices efficient?

- ◆ Is the system fair to all participants?

It is sometimes argued that these criteria can only be fully tested over a period of many years. However, I believe we have now had sufficient experience with the new system in England and Wales to judge that—provided the right market mechanisms are put in place, such as an efficient pooling system—the full liberalization model does work and does produce the desired efficiency effects even though the full effects may take some years to achieve.

The one caveat is that competition in supply has yet to be rolled out to small customers. The prospects for effective competition are quite good, but should it not work out it would certainly be possible to continue with a regulated monopoly supply activity within the general context of a competitive market.

It is clear that the partial liberalization model, as set out above, produces only some of the possible efficiency gains. It is equally clear, however, that this model could be adapted to achieve further benefits. For example, a utility with a supply monopoly could be required not simply to go out to tender for new capacity, but to operate its generating activity at arm's length and to enter into contracts on a non-discriminatory basis for all its generation requirements, perhaps through a portfolio of contracts of different lengths. Utilities might well be concerned, in such a scenario, about stranded assets; but their assets would only be stranded if, partially written down, they could not compete with new capacity—and if uneconomic assets are protected against such competition, the customer must be the loser.

Conclusion

I have no wish to be prescriptive about the precise system that should be adopted in any one country. Reform must reflect the starting point in each country and the characteristics of the local industry, ownership patterns, and culture. However, I think there are some common principles that would help to achieve benefits whatever the starting point.

Chapter 4: A Competitive Electricity Market

- Competition in generation is both beneficial and attainable, and as a minimum all new capacity requirements should be put out to tender.

- Vertical disaggregation is desirable to promote competition and also to achieve cost transparency and hence encourage efficient pricing and resource allocation. As a minimum, the separation (unbundling) of generation from network operations is desirable.

- A pooling system is necessary to achieve day-to-day least cost operation, where generating plant is under different ownership.

- Regulatory systems—controlling pricing for networks and any other monopoly activities and ensuring non-discrimination between market participants—are essential and should be designed to give the best incentives for cost reduction and efficient pricing.

I believe that the U.S. is bound to move further down the competition road; market pressures can be obstructed for a time but not stopped. There will undoubtedly be opposition—few monopolists willingly give up their monopolies—but while some of today's monopolies may struggle to live successfully in a more competitive environment, the customer can only win.

CHAPTER 5

The Shape of Things to Come: A Competitive Market in the Electric Industry

B. Jeanine Hull
Vice President, Environmental and Regulatory Affairs
LG&E Power, Inc.

Is competition in the electric industry good? Even though competition and the movement of regulated industries into more competitive postures have been going on in this country since the early '70s, there has been little public discussion of what is happening in the electric industry and how the trend towards competition is affecting the ways electricity is delivered.[1] I think it is crucial that we get widespread understanding and public involvement and that we not get caught up in arcane discussions and miss the real global issues that David Freeman discusses in Chapter 1.

Competition should not be an end in itself. The transition to a more competitive market should be structured to ensure a sustainable energy future and other socially desirable goals. Competition will not automatically achieve these goals. As Peter Bradford points out in his paper in Chapter 1, our goals will be our "guiding principles" as we move forward into a more competitive environment.

LG&E Energy Corporation reorganized in November of 1993 in anticipation of the changes expected to occur in the electric industry.[2] LG&E is now divided into two specific lines of business; retail electric and

---

[1] For this reason, I agree very much with what Ralph Cavanagh at the Natural Resources Defense Council is doing in terms of educating the public, although on a substantive basis I do not agree with his analysis.

[2] LG&E Energy Corporation is a holding company for a franchise utility, gas and electric utility. It is Louisville Gas and Electric in the metropolitan Louisville, Kentucky area. The non-franchise energy business is LG&E Power, Inc. Each group owns generating assets.

wholesale electric operations. The wholesale electric business now operates all of our generating assets, those of the franchised utility as well as those of the independent side. The retail electric business is a customer of the wholesale electric business, and the wholesale electric business has integrated a marketing function. There is no change in the legal ownership or regulatory treatment of any of the generating, transmission, or distribution assets. The reorganization was undertaken to enable each business to focus on its customers and to maximize customer satisfaction.

In the electric business, what is a competitive market in electricity? In Chapter 1 David Freeman provides a scenario in which there is a competitive generation market and franchise utilities remain the only distributors. However, if customers see a large disparity between production costs and as-delivered prices, such as I believe exists now in some parts of this country, they will press for more choices in selecting retail providers. Such a sequence of events occurred in the natural gas business; when the gas bubble hit, production prices dropped but customer prices rose. Marketers exploited this price delta and a whole new competitive natural gas business was born.

Is transmission a true, natural monopoly? And would transmission be a natural monopoly if capacity rights were tradeable? If customers win direct access to producers what will the role of the distributing utility be, that of an aggregator? If so, is that role justified only when the aggregator adds value to the commodity and does something that the customer itself cannot do? Is it when an aggregator is a common carrier? In our attempts to answer these questions, it is important to remember that an aggregator can facilitate economies of scale through bulk purchases and that these can be very profitable, especially when linked with the opportunities provided by the "information superhighway" that may provide innumerable new services to customers.

The California Commission's order in April declared that there will be retail competition in California in 1996. I believe that order was based on a fundamental and perhaps fatal mistake. The order confuses a market capable of sustaining competition with a market that is already competitive and open. I also believe that California's proposal that utilities be able to recover "administratively-determined stranded costs" will be publicly opposed just like administratively-determined avoided costs.

Let us look at the state of competition today. One data point is cogenerator deferral rates. These rates are a response of the utility to the perceived

Chapter 5: The Shape of Things to Come

competition of a cogenerator. These rates allow the utility to compete with the cogenerator; however, the cogenerator has no basis on which to compete against the utility. An example of this is now occurring on Long Island, New York. The Long Island School Board is refusing to pay the Long Island Lighting Company's (LILCO's) rates and declining the New York Power Authority's (NYPA's) rates. Instead, the school board relies on independent generators at a savings it estimates to be $55 million over a three-year period. LILCO is, of course, suing the school board.

The same thing is happening to industrial customers who pursue these kinds of savings. Industrial and commercial customers are looking for discount rates in reaction to a perceived increase in competition. Municipal power authorities are looking for alternatives to the suppliers to which they have traditionally been bound. Each is being sued by an electric utility that asserts that the customer has a "mandatory purchase obligation" to that utility.

Powerview and other new types of information systems being developed by Entergy give retail customers the information on which and therefore the ability to decide how, where, and when to use electricity. The development of information systems that will give customers the ability to control their use of electricity is, I believe, the only efficient DSM program. In addition, the development of options, futures, and other derivatives increases liquidity in power supply markets. Making the electric markets behave like typical commodities facilitates the participation of many other market entities.

Other examples of the changes in our markets due to the perceived increase in competition are the battles among municipal electric companies, customer cooperatives, and investor-owned utilities—also condemnation from municipalization. Southern California Edison recently won the right to accelerate nuclear depreciation to position itself for competition. There is a Florida case in which an independent power company is suing a Florida utility under anti-trust laws for failure to provide transmission service.

What do these data tell us? I believe they signal, very strongly, that this market is preparing for competition but is not yet competitive. What will a more competitive market look like and what will it take to get there? I think it is important to remember that although we can set up the structure we cannot define the outcome, in spite of the desires of the state and federal regulatory commissions.

While we can set up a structure to accomplish what we want to accomplish, we cannot define exactly what the result will be or what form the outcome will take. I think the most critical item in achieving competition, whether you define it either as wholesale or as retail competition, is that the distributing entity (i.e., the utility or entity that currently has monopsony purchasing rights in a geographic region) must be indifferent to its source of supply. Until it is, we may see fitful starts of competition here and there, or what passes to be competition, but we will not have an open and fair market.

Freeman writes that one of the ways to achieve an open and fair market is through divestiture. If you are the sole purchaser and you have an interest in one of the entities that is supplying you, you are by definition more likely to favor that entity than another supplier. If you are the owner of a supply resource, you can submit the ancient law, "if thine eye offend thee, pluck it out!" As Freeman has said, if your ownership of generation assets offends you, you can pluck them out and fully disintegrate them—or, to quote Phil O'Connor, you can "devolve," which is so much more gentle.

But there are other ways to get there, perhaps faster, maybe a little bit more easily, and maybe almost as efficiently as divestiture. One is though the establishment of real performance-based rate making. I distinguish real performance-based rate making from fake kinds by the presence or absence of three very specific components. First, an external objective benchmark must have been established against which performance can be objectively measured. Second, the purchasing entity contacts with every selling entity—including an entity that is affiliated with it—on the same terms and conditions and with the same written and enforceable contract. Third, the purchasing entity is truly indifferent as to source, and the purchaser is not better off buying from an affiliated entity. We must eliminate rate-based rate making where returns are based almost solely on the investments made. That means no new generation should be added to the generation rate base.

I think we'll start seeing transmission much more of a common carrier, probably on a regional basis. I am encouraged by Freeman's concept of a New York Transmission Authority under which all power is moved over a disinterested owner. We will eliminate state and federal laws that prohibit certain entities from entering specific markets and reform state laws that establish franchises. I also think it is important that utilities not be used as invisible taxing authorities to achieve social goals. I believe

Chapter 5: The Shape of Things to Come

the social goals should be promoted, but the way we go about doing it should not unfairly burden one of the entities.

I believe that after the transition to a more competitive market place has taken place utilities may be relieved of the obligation to serve, except perhaps to disadvantaged customers who do not have a choice. We cannot, however, relieve the utilities of their obligation to serve until a competitive market exists. Until the customers are assured that such a market is up and functioning, we cannot eliminate the obligation to serve. If we do it before, it will be like expecting somebody to walk on a high wire for the first time without a net. We need to keep that net in place for a limited period of time until a competitive market is up and running.

Industrial customers must be allowed to organize their activities to maximize their efficiencies at each plant site and among plant sites. It is imperative that we as a country allow our industrial base to be as competitive as possible. I think we are going to see a competitive marketplace in which generation, transmission, and distribution are optimized over larger geographic regions than currently, and that optimization will occur within a single system.

One predictable result of increased competition is that competitors that we have never imagined will evolve and that different services, different delivery systems, and different technologies will be developed. Our experiment with qualifying facilities (QFs) under the Public Utility Regulatory Policy Act of 1978 (PURPA) has demonstrated this. Qfs commercialized new technologies such as circulating, fluidized beds and many environmental control devices. We will start seeing, I believe, more fuel cells and more photovoltaic (PV) cells—plus a lot of things that we cannot foresee—as technology is released and research and development becomes a critical component in our ability to compete. How do we address environmental concerns as we move toward a competitive market? One of the effects of commoditizing electricity is the elimination of "environmental externalities." The energy supplier will very quickly internalize all of these social costs. Externalities could only exist in a rigidly price-regulated system. Furthermore, what we see very clearly in looking at the Qfs is that a significant technology advance occurred that enabled new technologies and new entities to "compete" in this new marketplace. We will see a leaner, meaner, regulatory function. I agree whole-heartedly with Peter Bradford that we will not see regulatory functions go away; we will see them focused on different parts of the business. They will be just as important and just as meaningful, but very different.

We need to redefine the boundary between state and federal jurisdictions. Intrastate and interstate distinctions are no longer viable. Retail and wholesale distinctions will also be redefined in a more competitive marketplace. There is, however, a difference between what the state regulators need to oversee and what the federal responsibilities will be. We also need to recognize that we will be in a regional marketplace. I envision the United States broken down into three basic regional markets. Those regional markets will have to be able to function in a coordinated fashion in terms of planning transmission and of moving power from one state to another without the artificial barriers that exist now between states.

Is all of this possible? Absolutely, but we must be careful in defining our goals. The goals, such as a competitive market, are important and so are the means of getting there. We must be careful that our means do not defeat our ends.

CHAPTER 6

Designing an Efficient, Competitive Electric Transmission Policy and Access Market Model

Dr. William W. Hogan
Thornton Bradshaw Professor of Public Policy and Management
John F. Kennedy School of Government, Harvard University

In 1992, Congress passed the Energy Policy Act, which among other things mandated steps that will lead to the creation of a competitive wholesale power market. This law and the policies that follow present a fundamental series of questions: what are the basic objectives of an efficient competitive wholesale electricity market, and how can those objectives be met?

When describing a market that includes competitive elements, it is useful to begin by listing a set of criteria that help to discriminate between different market designs. This paper explores such criteria and develops one possible market structure to meet these criteria.

Objective: Design a Market Structure that Includes Competitive Components

Some who propose introducing competition into the wholesale electricity market assert that it is not necessary to "design" a competitive market—simply to remove the constraints, and the market will evolve from there. The present discussion, however, starts with the assumption that a completely unfettered electricity market is neither possible nor desirable. Regulation of essential facilities, for example, will still be required. Hence, the model discussed below assumes that we will have to think carefully about how to design the competitive and regulated parts of the market structure in order to achieve that objective without compromising other objectives. An efficient design for an open-access wholesale electricity market must balance a number of competing objectives.

Reliability. Engineering limitations on operations must be respected, to preserve the stability and security of the electric power grid.

57

Least-cost Dispatch. The benefits of least-cost dispatch should be preserved, with appropriate pricing for the various services needed to support the associated power transmission.

Open access. This is now the law for wholesale transactions. Commercial functions must provide non-discriminatory, comparable access in the competitive sectors.

Revenue adequacy. The pricing framework should make possible the recovery of legitimate costs incurred in the creation of the existing transmission system.

Efficient trading and investment. Short-run trading should be economically efficient and compatible with long-run efficient investments and associated financial instruments.

Practicality. Transition steps should respect the remaining regulated components of the system.

Possible Structures:
Pooled and Bilateral Approaches to Market Operation

The dominant proposals for wholesale market structure and how it should operate have two broad themes on how we can organize the market. The first is a bilateral market, which operates primarily through individual arms-length deals arranged by the buyer and seller. This is what we generally consider the natural gas market. The other alternative is a "pool" market system, with some form of coordinated dispatch in order to achieve least cost under system security constraints and all of the other interactions that take place in short-term operations.

While it might be useful to compare the merits of these methods, the present discussion simply makes an assertion and then develops the implications of that assertion for a competitive wholesale market. The assertion is that, in the end, the U.S. will be best off with something approximating a pool-based market, like the system currently in operation in the U.K. This system can accommodate any desirable bilateral transactions. Hence, the pool-based market is fully comfortable with bilateral contracts and negotiations. As should be expected, there are aspects of the U.K. system that can be improved. The British have done a great job, but there are a few design details needing modification to address specific problems in the U.S.

Chapter 6: Designing an Efficient, Competitive Model

What type of wholesale market are we trying to achieve? At one end of the system we have many competing generators; at the other end of the system we have regulated distribution companies that provide services to final customers; and in the middle is the transmission operation, which is the critical essential facility for this market.

In this market, transmission can be separated conceptually into two different functions. One is the "Gridco," which is the group that builds and maintains the wires. This is an important function and having access to the wires is essential, but focussing solely on the wires is wrong.[1] The critical essential facility in this function is the dispatch function operated by the "Poolco," which does more than give access to the wires. The pool provides load following and reactive support, back-up supply, excess power sales, and all the other functions required in real operations in a real market. Until everybody has equal access to all of these services on the same basis, there won't be a level playing field. It won't be long before the Federal Energy Regulatory Commission's comparability standards embrace this requirement, much as they do in the natural gas business today. One way of solving the problem of equal service access is to have everybody integrated in pooled dispatch.

The assumptions to be made for the operation of that pool are essentially the same assumptions that apply in the British system. Electricity is treated as a commodity over a short period of time. There is a bid structure that the various participants in the markets can use to deal with all of the complexities of start-up costs, minimum and maximum outputs of the plants, and so on, just as they do in Britain. With this information, the system operator can provide a least-cost dispatch, along the lines of what is already done in the power pools in New York, New England, and the Pennsylvania-New Jersey-Maryland complex. Ancillary services can be handled through usage charges in this pool, and we can create a financial settlement system just as they have in England. This settlement system will be different from its counterparts in the U.S. today in that it will reflect the changes required by the competitive market.

---

[1] See Ruff, Larry. "Stop Wheeling and Start Dealing: Resolving the Transmission Dilemma. *Electricity Journal* 7(5), June 1994: 24–43.

The Electric Industry in Transition

Pricing Mechanisms: Electric Power Pricing

As a first approximation, we can describe the British power pool as a pricing system based on marginal cost. When the demand is low at 2 a.m., the operator runs the cheapest plants and charges their associated marginal costs. Later in the day when demand rises, the operator runs more expensive plants, charges this new, higher marginal cost to everybody, and pays the new marginal cost to all of the generators. Everyone faces the market price, not their costs of generation—an important feature of a competitive market. Later in the day, when the demand gets even higher, the price rises to an even higher level, and so on.

This pricing approach exists in the U.K. pool today. The dispatch system uses least-cost dispatch algorithms, much like the dispatch system uses least-cost algorithms in the New England and New York power pools. The critical difference is that, in the New England and New York power pools, the operators use split-savings methodologies and other approximations in order to parcel out the money after the fact. They don't apply the marginal cost pricing that would be required in a competitive market. The principal disadvantage of the split-savings system is that it produces different prices for everybody in the system. Any time there is a single commodity selling for different prices, it creates incentives for arbitrage. Everybody wants to go around the system and do separate deals. If they are allowed to do so, the pool breaks down. It is a fundamental principle of economics that for a homogeneous commodity there has to be one market price, and that the price ultimately should be the marginal cost. With the advent of competition, this pricing reform will migrate across the Atlantic.

In Britain, of course, the short-term market is not the only focus of activity. There is also a long-term contract market that provides hedges against the volatility of short-term spot market prices. The long-term contracts are all written relative to the spot prices that appear in the pool. All of the transactions that take place reflect what really happens—all of the power goes into the grid, is taken out of the grid, is charged at the short-term operating cost. Generators and customers can sign long-term contracts to hedge against the spot prices. The important and interesting feature of the British system is that those long-term contracts are made in a completely unregulated and private market that provides customers with real choice without affecting the pool dispatch.

The contracts that develop in the British system are long-term contracts that allow people to organize their own economics as they choose. To

Chapter 6: Designing an Efficient, Competitive Model

view the British system as having a purely short-term focus on short-term prices is to ignore most of the operations of the marketplace. The short-term system simply provides the necessary information and transparency to support really efficient long-term contracting and open access. The U.K. has a good system and it works well.

Pricing Mechanisms: Transmission Pricing

One thing that the British system does not do well, that we have an opportunity to fix here, is short-term transmission pricing. The British pricing system works under an assumption that all of the transactions are taking place at the same location, and that dispatch is unconstrained by any characteristics of the grid. The pool calculates the "system marginal price" under this assumption. Later, the pool makes many adjustments needed to recognize that in reality some plants are forced to run and some plants are excluded from use as a result of constraints in the transmission system. The resulting increased costs are spread around to all customers. This solution is imperfect and presents many problems in the form of poor incentives and cost shifting.

There is available a natural and simple solution to those problems that was developed at MIT by Fred Schweppe and his colleagues. The alternative approach takes into account the fact that electricity at different locations may have different prices. There is not just one marginal cost for everywhere in England—or in New England—there may be a different marginal cost price at every location. Furthermore, the price at every location can be calculated in a straightforward way that is consistent with least-cost dispatch. We in the U.S. can do everything that they in the U.K. are doing, but instead of having the trading taking place against one price for the whole nation, we would have one price for each location, with those prices linked in a way defined by the transmission constraints. Once we obtain those prices at every location, the cost of transmission from one place to another simply becomes the difference in the marginal cost at the different locations. If I sold a megawatt into the grid at one price and bought a megawatt elsewhere at another, then the difference in prices would be the cost of transmission. The difference in locational prices would define the opportunity cost of using the transmission system associated with that transmission in the short run.

The differences in those prices would include a little bit of power lost in transmission, but most importantly the transmission cost would include the difference in the marginal generation cost at the two locations. This difference would be the congestion cost caused by constraints in the transmission grid. This congestion rental can be quite large. It is not hard to find real examples where the congestion cost is of the same order of magnitude as the cost of generation itself; it would not be unusual. Losses might be relatively small, but transmission congestion can impose large costs at the margin.

If the pool charges marginal cost prices and collects these congestion rentals, there will be a substantial revenue flowing into the pool. What should we do with the revenue? We could create a set of transmission capacity rights in the system. These transmission capacity rights are not the rights to actually deliver power from point A to point B because, as everybody knows, the system operator might not be able to deliver the power because of other congestion that had arisen in the system. However, we can define the transmission capacity right as the right to collect the congestion rentals between the two locations.

Without going through all of the arithmetic here, note that in essence collecting congestion rentals between two points is economically equivalent to leasing back the physical capacity right and letting the right be perfectly traded in the market.[2] Market participants need not go through the exercise of leasing because the congestion rent is calculated automatically as part of the pool short-term marginal cost pricing. Hence we can avoid the very cumbersome and complicated market required every five minutes when everybody is trading their capacity rights back and forth. The trades occur automatically in the short-term dispatch and pricing. The capacity rights go to those who are paying the fixed charges for the grid infrastructure. Their right is to collect these congestion rentals. All the pool does is to charge the marginal cost to everybody who uses the system and pay congestion rentals to the people who have the capacity rights.

Hence there are two types of long-term contracts needed to integrate the short-term and long-term electricity markets. In England there are already long-term power contracts, which are entered into bilaterally by suppliers and customers. Additionally, in order to accommodate transmission congestion, we need long-term transmission "contracts". Unlike the power contracts, these transmission contracts have to be operated and managed by the pool, because only the pool has the information

---

[2] See Hogan, W. "Contract Networks for Electric Power Transmission." *Journal of Regulatory Economics* 1992(4): 211–242.

Chapter 6: Designing an Efficient, Competitive Model

about the congestion rentals that need to be rebated. There is no great complexity to creating this transmission pricing and rights part of the system, and one of the enormous advantages of such an innovation would be that all of the problems of loop flow and changing capacity for the system could be handled automatically in the dispatch. Most of the problems that have been causing such concern about where the power is actually flowing, and whose power is flowing over which lines, and so on, are handled automatically under the suggested pricing and dispatch system with the associated power and transmission contracts.

Adoption of this scheme of transmission pricing means that we would end up with a new system for cost recovery of transmission charges. We must address the existing system cost and the existing transmission rights. This will be a difficult and contentious allocation process because we haven't been explicit about what those rights have been in the past. We must, however, be explicit about transmission rights in the future. New transmission investments and capacity rights can be handled quite simply by requiring people to agree to pay for them. Trans Power of New Zealand embraced this policy: no new transmission investments would be made until the customers signed the contracts to pay the fixed charges. The transmission lines themselves, with their poles and capacitors, might be owned by someone completely different. The open-access system separates ownership and usage. The transmission capacity rights can be traded in the marketplace, and all the variable operating costs are handled through the marginal-cost prices. The spot prices give a well-defined measure of opportunity cost, continually, and through the capacity rights we would have a simple mechanism for protecting native load users without compromising open access.

Conclusion: We *Can* Design a System that Meets Our Objectives for a Competitive Power Market

The transition problems associated with this or any other change in market structure are a subject for another discussion. The purpose of this paper has merely been to assert that, if we were to adopt a system like the one described—a pool-based system—it would be possible to create a market framework that would meet the objectives set out above. Reliability criteria will be met automatically by preserving the existing dispatch criteria and dispatch procedures to make sure the wires don't fall down. We would preserve least-cost dispatch that now exists in the New England or the New York power pools. Open access is automatic—everybody is there on the same level playing field. By definition, with open access anyone could connect to the wires, and anyone could buy and sell power

The Electric Industry in Transition

at short-term marginal costs just like anybody else. If market participants want to contract around the short-term market for the long run and trade rights and make investments, they can do that too, but this contracting would be an unregulated part of the business. For overall revenue adequacy of the transmission infrastructure we rely, not on the short-term payments where transmission used to pay for the system, but on fixed charges. The transmission capacity rights to collect congestion rentals can be traded.

Finally, this system is practical. This system is easier and simpler to operate than the current system of economic dispatch with split savings after the fact, yet the proposed system is compatible with a competitive market in generation.

Part 3

The Electric Industry in Transition

INTRODUCTION

What are the Costs and Benefits of Market Efficiency and to Whom?

Charles R. Guinn
Deputy Commissioner for Policy Analysis and Planning
New York State Energy Office

The reason usually cited for opening markets to greater competition in the electric industry is the promise of greater economic efficiency and hence lower electricity costs. The reason often cited for slowing the pace of competition, especially at the retail level, is the potential for increased costs to residential or captive customers. The basic question which needs to be addressed is whether we are playing a "zero-sum game" reallocating costs from large to small customers through deregulation or really striving to reduce the costs of the industry through market forces.

The public's desire for and support of regulatory reform allowing greater competition will likely be determined by its belief in whether the change will be mostly a reallocation or reduction in costs. If some of the customers are likely to be winners and some losers, deregulation will be difficult and dependent in part on the political influence of each side. However, if all customers can be expected to benefit and many by a significant amount, the path will be smooth and defining an orderly pace of change will be the challenge. Another question to be addressed is the temporal distribution of benefits and costs. Short-term lower costs may lead to greater long-term costs.

The authors in this section do not agree on the need for competition or the current degree of it in the electricity industry. Nor do they cite likely costs and benefits to differing classes or to society in support of their positions. However, there is a common theme in their papers—that a managed transition to a more competitive electric industry is a responsibility of government that must be carried out carefully. Another common thread is that establishing competitive markets in wholesale generation will, both in the immediate future and in the long term, be the logical first step in making the industry's market more competitive.

Armond Cohen argues that an environmentally sustainable electric industry is or should be our primary social goal and that movement to more competitive electricity marketing methods, including retail wheeling, would greatly endanger our ability to attain societal goals. He argues that retail competition (like that in the United Kingdom) would lead to an elimination of DSM programs; procurement of clean, renewable resources; and research and development of new technologies.

Cohen challenges the proponents of retail wheeling to document the benefits of retail wheeling beyond those that would likely also be available from wholesale competition. He argues that retail wheeling cannot reduce embedded costs, only reallocate them, and that the likely impact of retail wheeling will be to raise prices—especially in the long term.

Cohen's prescription is to focus on practical steps to pursue wholesale competition and ignore the theology of markets, pursue embedded cost reductions directly through asset repricing, and incorporate environmental risks into the integrated resource planning process.

Philip R. O'Conner's primary point is that competition is now here to stay, and that information rather than government rules should be the regulator of choice. Information as a regulator of market behavior would enhance choice and value to the customer and lead to a better overall utilization of assets. Thus society as a whole would benefit through greater choice (value added) and lower costs.

O'Connor offers a method he calls "progressive choice" to encourage greater competition without having to wait for the retail wheeling debate to be concluded. The progressive choice model would allow the benefits of competition to percolate to all classes, end micromanagement of the industry by regulators, address transitional issues associated with stranded investment, and rely upon new communication and information technologies to enable customers to exercise greater choice and utilities to tailor services for specialized needs. A belief that at some time in the future most customers will receive and base purchase decisions on real-time price information is at the heart of the progressive choice model.

P. Chrisman Iribe espouses a belief that the potential benefits to consumers outweigh the costs of increased competition. He argues that a truly competitive wholesale generation market will sort out the winners and losers among generation providers and in turn pass the benefits of competition along to the ultimate customers. A basic premise of his argument is that generation companies will need to add value to the product by managing risks for customers.

Iribe sees a vision of the future generation business under wholesale competition in which merchant plants sell both short- and long-term contracts, generators make significant sales to power brokers and marketers, generation is unregulated and unbundled from transmission and distribution, the independent power industry is consolidated, there are joint ventures on specific merchant plants, and utilities spin off generation.

In summary, the session panelists agree that the winds of change are blowing and that while we may not be able to direct the winds, we can adjust the sails. We hope the course we choose for introducing and accommodating competition in the electricity industry can follow a variety of paths and provide real benefits to all customers and society.

CHAPTER 7

The Political Economy of Retail Wheeling, or How to Not Re-Fight the Last War

Armond Cohen
Senior Attorney and Energy Project Director
Conservation Law Foundation

Disparities in utility rates—observably the result of poor supply-side resource planning—have been small before and will be small once again. Retail wheeling's promise of short-run gains for a few would, ironically, destroy integrated resource processes in place today that guard against a report of yesterday's planning mistakes.

"Politics," quipped Groucho Marx, "is the art of looking for trouble, finding it everywhere, diagnosing it incorrectly, and applying the wrong remedies." The estimable comedian might have been describing the current "retail wheeling" debate in the United States.

For all the ink it has consumed and passion it has aroused, the U.S. retail wheeling debate is still remarkably devoid of the key elements of a rational policy discussion: a reasoned definition of the "problem," an analysis of its causes, and an examination of alternative solutions. This article attempts to sort our the underlying causes of and interests at stake in the retail wheeling debate and assess the validity of the assertions of retail wheeling proponents.

We conclude first that support for retail wheeling is in significant part a reaction to the utility rate differential resulting from the ambitious capacity construction programs (most but not all of them nuclear) of the 1970s and 1980s. Ironically, these construction programs, and resultant cost overhangs, were considerably smaller or nonexistent where there existed integrated planning of the type that many retail wheeling proponents decry. Driven by the desire to escape these embedded capacity costs, retail wheeling proponents are essentially still fighting the last war—the war over nuclear imprudence. Sunk cost shifting, not prospective cost-savings, is their end. Political muscle is their means.

Even if this effort actually results in significantly reduced rates to wheeling customers in the short term (and there are many reasons to believe that in any fair system, it will not), over the long term the energy policy implications of retail wheeling are problematic indeed—a feature it shares with more thoughtful and intellectually honest U.K. industry restructuring proposals. Specifically, the replacement of an electric power planning framework with one dominated by bilateral contract is likely to be a poor fit in a world where the dominant drivers of electric power policy are likely to be environmental. It also runs counter to the direction in which technology is driving power systems—toward smaller scale, distributed generation requiring more system coordination and planning, not less. Finally, it fails the test of political legitimacy that will be critical as environmental debate over the shape of the power system continues and intensifies.

In short, retail wheeling is a troubling answer to a mis-diagnosis of yesterday's problem. We believe that a variety of other policies offer most of the benefits and few of the risks of retail wheeling. These include aggressive wholesale competition, judicious pruning of uneconomic capacity, and serious incorporation of environmental risk considerations into utility planning and regulation.

Retail Wheeling: Imprudence Litigation by Other Means?

The now-familiar case for retail wheeling rests on a chain of assertions that are rarely scrutinized: that electric rates are a significant factor in U.S. (or state) industrial competitiveness; that high industrial rates are due to the absence of retail competition, industrial-to-residential cross-subsidies, and the imposition of integrated resource planning requirements; and that the dissolution of the retail franchise will result in significantly lower rates and therefore in high industrial employment.[1] While all of these assertions are shaky and unsupported[2], here we focus on the assertion that current high embedded rates can be explained by the absence of retail competition, cross-subsidies, and the development of integrated resource planning requirements.

---

[1] *See,* e.g., J. Anderson, "The Competitive Sourcing of Retail Electric Power: An Idea Who's [sic] Time Has (Finally) Come," (Presentation to Utility Directors' Workshop, Williamsburg, Va.) (Sept. 10, 1993); J. Anderson, "Wheel in Cheaper Power from Quebec," *Providence Journal* (Mar. 29, 1993); "PUC Chief says buying cheap Canadian power wouldn't drop local costs," *Providence Journal-Bulletin* (May 12, 1993, at D10).

Chapter 7: The Political Economy of Retail Wheeling

U.S. Retail Electric Rates 1970-93

In many jurisdictions where it has been discussed, the practical impetus for retail wheeling is the differential in retail electric rates between utilities in selected jurisdictions—especially the differential in industrial customer rates in neighboring utility service territories. Table 1 represents a typical inter-utility comparison, which is derived from a list compiled by Charles Studness.[3]

Thus, an industrial customer of one of the depicted high-cost utilities could theoretically save, on average, about 34 percent if it were served by the neighboring low-cost utility. Such a customer is likely to be unhappy that other customers in the same region, possibly competitors, are able to buy electricity at such a discount relative to the rate it pays. Hence the clamor for retail wheeling among some (but by no means all or a majority of) industrial customers.

This leads us to wonder whether these rate differentials have existed over time or whether they are a more recent phenomenon. To answer this question, we first examined the 1970 retail industrial rates of these utilities (see Table 1).

While the rates in the table are noticeably lower than rates today, the key point is that the differentials are much smaller, both absolutely and relatively.

---

[2] Electric costs typically represent roughly 2% of total manufacturing costs in the United States, hardly a decisive factor in manufacturing profitability and industrial employment (MSB calculation, based on U.S. Census of Manufacturers, 1987). Even if these costs were significant and could be reduced substantially through retail wheeling, the net job creation effect is likely to be negative, since larger industrial customers would be far more likely to "benefit" and transfer embedded capacity costs to small business. Yet it is small businesses with fewer than 20 employees which account for 80% of new jobs created during the period 1987-92; by contrast, large firms have had modest or negative job-creation records over the same period. Cognetics, Inc., "Who's Creating Jobs?" (Cambridge, Ma. 1993). It is hard to see how shifting electric costs from job-shaders to job-producers will increase employment.

[3] C. Studness, "The Geography of Electric Rates," *Public Utilities Fortnightly*, Oct. 1, 1993, at 36.

The Electric Industry in Transition

Table 7-1. 1993 vs. 1970 Retail Industrial Rates; Percent Nuclear Power in Generation Mix

High-cost Utility	1993 Rate[a] (Cents)	1970 Rate[b] (Cents)	Generation Mix[c] (% Nuclear)
Doe Inc.	6.4	1.3	30
Philadelphia Electric	7.8	1.2	58
Long Island Lighting	11.9	2.0	9
Illinois Power Company	4.7	1.5	22
Ohio Edison	6.2	1.4	25
NIPSCO Inc.	4.9	1.4	0
DPL Holdings	4.8	1.2	0
Centerior Energy	6.6	1.3	56
Commonwealth Edison	6.2	1.6	83
General Public Utilities	6.0	1.2	23
Pub. Service New Mexico	6.2	1.1	31
Orange and Rockland	8.3	0.8	49
Entergy	5.9	0.8	49
IES Industries	4.6	1.8	24
Detroit Edison	6.8	1.1	16
Average	6.5	1.4	28

Low-cost Utility	1993 Rate[a] (Cents)	1970 Rate[b] (Cents)	Generation Mix[c] (% Nuclear)
American Electric Power	3.8	0.8	6
Potomac Electric Power	5.5	1.7	0
Pennsylvania Power and Light	6.0	1.5	31
IPALCO Inc.	4.2	1.3	0
LG&E Energy	3.8	0.9	0
PSI Resources Inc.	3.5	1.1	0
KU Energy	3.4	1.6	0
Cincinnati Gas and Electric	4.4	1.4	0
Wisconsin Energy	3.9	1.7	31
Alleghany Power System	4.2	0.9	0
Southwestern Public Service	3.6	1.3	0
Delmarva Power and Light	3.8	1.4	0
Empire District Electric	3.8	1.4	0
Northern States Power	4.4	1.7	28
CMS Energy	5.1	1.3	14
Average	4.3	1.3	9

a. *Source: Prudential Securities, Electric Utilities: Competitive—First Study. September 24, 1993.*
b. *Source: Moody's Public Utility Manual 1971.*
c. *Source: The Value Line Investment Survey.*

This table shows why there was little interest in retail wheeling in the past. In 1970 an industrial customer switching from a high-cost utility to a low-cost utility would save only about 7 percent, in contrast to the 34 percent average savings available to these customers today. Industrial customer rate differentials between utilities in the 1970s tended to be much smaller than they are today.

While some of the clamor for retail wheeling has cited the differential between current IPP marginal costs and utility rates, in addition to inter-utility rates, the likely medium-term retail wheeling scenarios involve inter-utility competition, until much more far-reaching deregulation, divestiture and spot market mechanisms are worked out, as in the U.K. In any event, comparisons between nuclear utility and IPP rates are likely to follow the same trends and patterns as comparisons between high- and low-cost utility rates, described below.

Next we examine what happened over the past two decades to cause the high-cost group's rates to escalate so dramatically relative to the low-cost group's rates to generate today's huge rate differentials.

The Cause of Rate Differentials: Poor Supply-side Planning

The major cause of the large rate differentials today is clearly related to poor planning on the part of the high-cost utilities. Many of these companies made big bets on nuclear power, encouraged by what was then a relatively "hands off" state regulatory environment, their belief in ascending economics of scale, and the absence of requirements for wholesale competition. In general, the companies lost those bets.

In the late 1970s and 1980s the cost of building nuclear plants escalated dramatically. Utilities with ambitious nuclear construction programs tended to see their rates increase substantially. Those that avoided the nuclear option or had a more balanced portfolio of resources, including DSM, saw rates increase much more slowly. Given this backdrop one might expect that the high-cost utilities have more nuclear generation. They do. In fact these companies have, on average, over three times as much nuclear generation as do their low-cost counterparts (see Table 1).

Table 1 actually understates the effect of the nuclear problems of the late 1970s and 1980s. Some low-cost utilities that have relatively high amounts of nuclear generation (e.g., Wisconsin Energy) built their plants in the late 1960s and early 1970s before the staggering construction cost escalations and construction delays. And some high-cost utilities (e.g.,

Long Island Lighting) spent substantial amounts on nuclear construction during the late 1970s and early 1980s, but failed to add the cost of the nuclear plant to the rate base. In spite of these potentially mitigating effects, nuclear construction problems stand out as a root cause of the rate differences.

To be sure, nuclear investments are not solely responsible for the currently observed rate differentials. Expensive IPP contracts based on administratively determined avoided costs (often based on proxy nuclear costs) rather than wholesale bidding also play a role in many jurisdictions. In addition, low oil and gas prices exacerbate these differences.

The key point is that the conventional pre-wholesale competition electric supply-side planning practices predating integrated resource planning (IRP), not demand-side management or inclusion of externality values in planning, contributed most significantly to the distorted rate picture we see today. In fact it is highly likely that if good integrated resources planning—such as Wisconsin Energy and New England Electric Systems did—had been done, if a regime of market-based wholesale competition had been in place, or if environmental regulatory risks had been part of regulatory evaluations, many of the expensive nuclear plants would have never been built and rate differentials would have been substantially moderated.[4]

The Current Rate Differentials Are Not Due to Cross Subsidies

Some retail wheeling advocates claim that rate differentials are due to cross subsidization of residential customers by industrial customers of the high-cost utilities.[5] The evidence does not support such a claim. The

---

[4] After constructing a low-cost plant (Point Beach) in the early 1970s, Wisconsin Electric proposed to build additional units in the mid to late 1970s. Due to planning uncertainties, however, the Wisconsin Public Service Commission denied these requests. Likewise New England Electric System, during the late 1970s and early 1980s, consciously scaled back its nuclear commitments and undertook a commitment to fuel diversification and conservation. It currently has rates considerably below the New England average.

[5] It is not clear that industrial customers actually subsidize residential customers. In fact the subsidization may run the other way. For a discussion of the potential for subsidization of industrial customers by residential customers by residential customers, see David Schoengold, "Allocating the Cost of Generating Capacity: A Discussion Paper on the Question of Interclass Subsidies (Manuscript, MSB Energy Associates, Inc.)," (Nov. 1993).

ratio of residential rates to industrial rates for the high-cost utilities is about the same as it is for the low-cost utilities (see Table 2). If any difference exists, it is that the high-cost utilities have a slightly higher ratio of residential to industrial rates. In other words, industrial customers of high-cost utilities have slightly greater discounts relative to residential rates than do the industrial customers of the low-cost utilities. Therefore if the suggested subsidies do exist and are corrected for all utilities, it would make the current rate differentials larger, not smaller.

The truth is that both residential and industrial rates of the high-cost utilities are high relative to their neighbors' rates because these utilities invested in expensive, large, and risky supply-side resources supported by a regulatory environment that showed little interest in supply-side modularity, environmental risk, or wholesale competition. Supply-side resource costs have been (and, as noted below, will continue to be) the primary driver of rates.

Given this background, it is not surprising that some industrial customers are utilizing whatever arguments and pressure they can muster—including the threat of retail wheeling—to get more rate relief from high embedded supply costs than they got through the nuclear imprudence litigation of the 1970s and 1980s. Despite the fact that in some regions of the country industrial interests supported large-capacity construction projects to provide necessary reliability[6], the response of many retail wheeling proponents is essentially that someone else (utility shareholders or other ratepayers) should pay for costly power acquired on their behalf now that the bill has come due.[7] "Stranded investments," and cost shifting to inelastic customers is therefore not an incidental part of the retail wheeling proponents' agenda. It is the primary object of that agenda.

---

[6] See., e.g. THE NEW ENGLAND COUNCIL, REPORT ON THE REGIONS ENERGY FUTURE (Nov. 1990); Mass. DPU Docket 87–169, Pre-Filed Testimony of Karl Christ on Behalf of the Energy Consortium; R. Buck, "Our Need for Power Plants," *Providence Journal* (May 14, 1989).

[7] "PUC chief says buying cheap Canadian power wouldn't drop local costs," *Providence Journal-Bulletin* (May 12, 1993). [Retail-wheeling advocate Edward] Burke questioned why the PUC is so interested in protecting private utility companies that, he said, made bad decisions to invest in expensive nuclear plants and now charge Rhode Island industries the nation's highest electricity rates."

The Electric Industry in Transition

As in all public policy discussions, through, there is a danger in rearview vision, and making prospective policy to solve yesterday's problems. Specifically, we argue, the retail wheeling argument confuses transitory cost bulges with long-term differentials, ignores necessary regulatory arrangements that will diminish the attractiveness of system bypass, and badly misreads the terrain that lies ahead.

The Near-Term: Diminishing Returns in an Equitable Retail Wheeling Regime?

The premise of retail wheeling proponents is that substantial rate differentials among neighboring utilities will persist, and that therefore the freedom to "shop around" is worth the price. However, there are a number of reasons to believe that these differentials will not persist, or that they will not translate into the predicted bargains.

Average Price/Marginal Cost Differentials Will Narrow

First, even if there are no changes in ratemaking, utilities are not likely to repeat the "nuclear" mistake. Utilities that overbuilt have not in general been able to spare their shareholders from the consequences of their poor planning and consequently a new regulatory bargain has emerged (see Table 2). Utility managers have learned that their choice of new generation will be scrutinized against a wholesale market-based benchmark, that there will be close regulatory scrutiny of the selected generation portfolio, and that plant modularity and environmental risks will be key elements in that scrutiny. In this connection, claims that utility DSM programs represent the next iteration of the "nuclear syndrome"[8] are not credible. DSM advocates and skeptics alike are ensuring that DSM programs are thoroughly evaluated and scrutinized for cost effectiveness; the programs have generally been accompanied by cost recovery penalties if the programs fail to deliver; and such programs, unlike generating plant, lend themselves to swift revision or cancellation where shown not to be cost effective.

Likewise, new supply-side capacity commitments are likely to receive enhanced scrutiny of their modularity, riskiness, and timeliness.

---

[8] *See, e.g.* B. Black and R. Pierce, "The Choice Between Markets and Central Planning in Regulating the U.S. Electricity Industry," 93 Colum. L.J. 1431–32 (1993).

Chapter 7: The Political Economy of Retail Wheeling

Table 7-2. Ratio of Residential Rates to Industrial Rates[a] and Stock Price Change (1970–1990)[b]

High-cost Utility	Ratio	% Change
Doe Inc.	2.0	-1
Philadelphia Electric	1.7	-19
Long Island Lighting	1.3	-11
Illinois Power Company	2.2	-53
Ohio Edison	1.7	-14
NIPSCO Inc.	2.1	17
DPL Holdings	1.6	17
Centerior Energy	1.7	NA
Commonwealth Edison	1.5	-3
General Public Utilities	1.5	111
Pub. Service New Mexico	1.5	-43
Orange and Rockland	1.5	51
Entergy	1.8	-10
IES Industries	2.1	NA
Detroit Edison	1.4	37
Average	1.7	+2

Low-cost Utility	Ratio	% Change
American Electric Power	1.7	9
Potomac Electric Power	1.3	214
Pennsylvania Power and Light	1.4	86
IPALCO Inc.	1.4	110
LG&E Energy	1.6	17
PSI Resources Inc.	1.7	-37
KU Energy	1.3	49
Cincinnati Gas and Electric	1.5	22
Wisconsin Energy	1.8	310
Alleghany Power System	1.7	87
Southwestern Public Service	1.7	156
Delmarva Power and Light	1.8	61
Empire District Electric	1.5	69
Northern States Power	1.6	190
CMS Energy	1.3	-7
Average	1.6	+89

a. *Source: Prudential Securities, Electric Utilities: Competitive—First Study. September 24, 1993.*
b. *Source: The Value Line Investment Study. 1970 stock priorities are not available for Centerior Energy or IES Industries due to their major corporate reorganizations in the 1990s. Note that over the last two decades, the stock prices of the high-cost utilities have on average essentially remained unchanged. The significant negative impact on shareholders becomes more apparent when compared to the near doubling of the stock prices for the low-cost group over the same period.*

The Electric Industry in Transition

Second, as the assets acquired during a period of expensive overbuilding are depreciated, the rate differentials will return to more normal levels. While the python's digestion of the nuclear rodent is likely to be slower than many would like—or slower than can be achieved through alternative political means—it is inevitable.[9]

Third, many of the utilities with low rates today (especially coal-based Midwestern and mid-Atlantic utilities) face large Clean Air Act compliance costs that will increase costs and rates on their system relative to more nuclear reliant systems.

Fourth, in markets such as the Northeast and Midwest, much of the presently observed price differential is an artifact of excess generating capacity. There are a number of factors, aside from the normal economic recovery pattern, that will tighten those markets. These factors include premature nuclear unit retirements (recently Wall Street analysts have predicted up to 25 such unit retirements within the next decade)[10] and premature retirement of existing fossil fuel units due to the cost of meeting Clean Air Act requirements.

System Price/Risk Realignment and Transition Arrangements May Negate a Substantial Portion of the Perceived Short-term Gains

Among industrial retail wheeling advocates with whom we have spoken, there seems to be a curious assumption that the system in which they seek to shop around will be largely unchanged except for their ability to get lower-cost generation. This assumption is wholly unsupported.

First, we agree with the assessment that "the obligation to serve that utilities traditionally have operated under must be ended if retail wheeling is to be introduced rationally," even for customers who elect not to

---

[9] The Maine PUC recently rejected contentions by Central Maine Power of "substantial and long-term" average price/marginal cost differentials, citing this point and many of the factors discussed in the text below. *See* Maine PUC, Investigation of Central Main Power Company's Resource Planning, Rate Structures, and Long-Term Avoided Cost, Docket No. 92–315, Feb. 18, 1994, at 26–27; 32–33.

[10] P.C. Parshley, et.al., "Should Investors be Concerned about Rising Nuclear Plant Decommissioning Costs?" Shearson Lehman Brothers, Jul. 6, 1993.

shop.[11] This has, in fact, been the rule adopted in the U.K. and Norway, where retail wheeling has been introduced as part of broad industry restructuring. The corollary of the demise of the obligation to serve is that customers who leave the system must return "under market-determined—not regulated—prices, terms and conditions."[12] It is anybody's guess how many U.S. industrial customers—particularly large ones with sensitive production processes—would prefer to take on the economic risks of a non-guaranteed power supply secured only by the contractual liability of retail power brokers of the ethical and political compunctions of a distant IPP utility, but it is highly likely that maintaining current levels of supply security will be difficult and come to such customers only at substantially higher levels of cost.

Second, a retail wheeling regime will inevitably require a more detailed and very different pricing regime for continued utility system support, which takes into account the costs and risks of providing various levels of capacity firmness, reserves, transmission support, balancing, voltage support and other services. These factors, too, are likely to erode some of the apparent generation bargain.[13]

Third, a retail wheeling regime would no doubt lubricate movement of substantial market-based reforms in transmission pricing. It would be hard for retail wheeling advocates and regulators to advocate and implement a system which forces a utility to "write down" the value of embedded generating assets to market levels without at the same time "writing up" the value of the transmission and distribution system to market levels. Conceivably such a "write-up" could offset entirely the initial "write-down" of generating assets.

Fourth, the write-down of nuclear generating assets resulting from retail wheeling is likely to have some peculiar system and price effects. If the U.K. experience holds, the market value of nuclear generation in a world

---

[11] Rodney Frame, "Characteristics of a 'Good' Retail Wheeling System" (presentation to Electric Utility Business Environment Conference, Denver, CO, Mar 16–17, 1993) at 6.

[12] Frame, *supra*, note 11.

[13] Frame, *supra*, note 11.

without the retail franchise is likely to be zero or negative due to nuclear operating risks and decommissioning costs. For some utilities, the book value of nuclear assets approaches or exceeds total utility equity, making bankruptcy a likely prospect. Will market players emerge to purchase and operate these zero-value assets in a safe and reliable manner, and assume the substantial unfounded unit decommissioning costs to boot? Again, if the U.K. experience holds, the answer is no.

Accordingly, strategies will need to be devised to address this market gap. The most likely political solutions are a federal bailout or all-customer "transition charges" to cover unfunded nuclear decommissioning and operation costs. In either case, wheeling customers will pay as customers or as taxpayers. Furthermore, if the plants are decommissioned on an accelerated basis, the disappearance of that capacity will, in some regions of the country, accelerate the convergence of embedded costs and market price discussed above.

Fifth, the dissolution of the retail franchise is highly likely to increase dramatically the risk of financing new generation (or refinancing existing generation) from whatever source. Downgrading of utility financial ratings is already occurring due to capital markets' perceptions of increased retail franchise risks within the electric utility industry.[14] Widespread retail wheeling would likely exacerbate this trend because of the increased risks it creates for utility investments and operations. It is folly to believe that marginal generation costs will not reflect this increased financial risk.

In short, retail wheeling will not be the cost-shifting bargain that its advocates expect it be, assuming that the above issues are dealt with up front and in a fair and rational manner. Apparently recognizing this fact, some retail wheeling advocates have argued that these issues are "mere details" that "we can talk about later" after customers are allowed retail wheeling. And some have dismissed the arguments above by contending that market-based pricing of the transmission system "def(ies) accurate identification measurement" and would encourage "monopoly rents";

---

[14] Daniel Scotto, "Weakness Confirmed: Moody's and S&P's Pronouncements Do Not Bode Well for the Electric Industry, *Fortnightly*, Dec. 1, 1993, at 34–36.

that the presumption of an existing obligation to serve is "not realistic" as a starting point; and that it is effectively "impossible" to establish a market-based rule for system "return rights."[15]

It is precisely this evasion of critical issues by retail wheeling advocates, and the consequent potential for piecemeal cost-shifting and political manipulation, which has led even staunchly "pro-market" power system economists to reject retail wheeling as a credible starting point. For example, Ruff states that in the United States, competition in electricity is being defined in terms of "wheeling" rather than in terms of the open pooling and transmission model outlined here. As a logically consistent statement of how an electricity system can combine effective competition with economic efficiency, the wheeling model is seriously deficient or even non-existent.[16]

Similarly, Joskow posits that "a regime that relies extensively on competition can work reasonably well if all of the right pieces are put in place at the outset." He adds, "Retail wheeling in the U.S. is likely to merge without all the right pieces in place and, as a result, will probably be costly and inequitable."[17]

The Promise of Prospective Real Cost Efficiencies Is Not Credible

We have argued that the motivation for current retail wheeling proposals is an attempt to redistribute the cost of pre-IRP, pre-wholesale competition supply mistakes—mostly nuclear. But some advocates argue that retail wheeling will have salutary prospective pro-efficiency effects, citing recent utility downsizings.

---

[15] *See, e.g.,* Anderson, *supra,* note 1, at 13–16.

[16] Larry E. Ruff, "Competitive Electricity Markets: The Theory and Its Application," Dec 1992, at 32 (forthcoming in a volume on electricity markets, M. Einhorn, ed. Kluwer Academic Publishing Co.).

[17] Paul L. Jaskow, "Emerging Conflicts Between Competition, Conservation and Environmental Policies in the Electric Power Industry," draft at 26 (keynote address for Public Utility Research Center conference on competition in regulated industries, U. of Fla., April 29–30, 1993).

On the supply-side, there are certainly likely to be prospective cost savings whenever a utility is well-focussed on obtaining the most competitive prices for new capacity. However, as we argue below, that incentive can be created more directly, and with far less damage to other interests, through a regime of wholesale competition in which, prior to acquiring new capacity, utilities must demonstrate through bidding or other competitive tests that they have obtained the best possible market deal. Retail wheeling is unlikely to achieve any additional generation cost efficiencies. No one has argued, as far as I am aware, that end-use customers can obtain new capacity less expensively than utilities.

Apart from generation, there could also be marginal efficiencies in administrative and general operations through downsizing or other restructuring. However, the effect of such gains here is limited: administrative and general expenses typically account for a small fraction—as little as 5% of the overall utility revenue requirement—and will in any event have to be borne in some form by customers, whatever their supply sources.[18] The principal component of utility rates will continue to be—as it has been historically—supply-side costs.

The most that can be said for the prospective impact of retail wheeling on supply costs is that it may permit a future shifting of costs away from customers with short- to medium-term contracts to generators who have made "bad bets" as defined by short- to medium-term markets, although it will not prevent the social expenditure of capital on those bets. In addition, as described below, the price of such prospective cost shifting is likely to be the radical shortening of system investment criteria and hence substantial foregone medium- to long-term efficiencies at the end-use and system level.

Summary

We have argued that even in the near term cost-shifting rate reductions to retail wheeling customers are likely to be significantly diminished by factors driving up system marginal costs, as well as from the realignment of system operations, risks and component pricing that would flow from any rational retail wheeling regime. We now turn to the long-run implications of retail wheeling and operations.

---

[18] Calculation by the Authors based on analysis of FERC Form 1 data for selected mid-Atlantic utilities.

Chapter 7: The Political Economy of Retail Wheeling

Long-run Impacts of Retail Wheeling: The Road Better Left Untaken

While retail wheeling is surely an opportunistic, near-term strategy to shift embedded capacity costs on the part of its proponents, it will have obviously long-term consequences for power system planning and regulation. This point has at least been acknowledged by those proposing or examining more far-reaching U.K.-style utility industry restructuring that includes dissolution of the retail franchises.

The emerging picture of such a world contains some obvious features. Generation and transmission construction decisions will be driven by near-term bilateral contract dynamics rather than by traditional planning concerns of medium- to long-term system stability, diversification, and fuel price risk or newer concerns of modularity, plan robustness and environmental regulatory risk. Lacking long-term customer/utility relationships, and driven by prices rather than costs, power system investments in energy efficiency will disappear entirely, as they have in the U.K. and Norway. Finally, environmental concerns, now reflected in a variety of ways through IRP processes and state facility-siting regulation, will be wholly externalized into the realm of taxation and end-of-the-stack regulation. In short, much of what we call "energy policy"—that is, the stuff that fills the pages of this journal—will be replaced or marginalized by the imperatives of bilateral contract.

While the broad implications of this post-retail wheeling world are addressed by other contributors to this issue, here we focus on three questions.

◆ Is such a world compatible with a sensible and cost-effective approach to energy-related environmental problems in the coming century?

◆ Is such a trend likely to fit with an emerging world of technological opportunity emphasizing economically efficient distributed generation?

◆ And does such a world satisfy the tests of political legitimacy and transparency?

Environmental Implications

We start by noting that the most critical economic and public policy challenges associated with the power system in the coming decades are likely to be environmental in nature. Current Clean Air Act compliance directives in the area of ozone smog and acid rain are just the beginning. A variety of additional power system-related environmental concerns are at the beginning or the middle of the regulatory pipeline and are likely to emerge with unexpected rapidity. These include, at a minimum, toxic air emissions, small particulate emissions (now believed to cause more annual U.S. fatalities than auto accidents), and evidence that existing ozone smog and local sulfate standards are insufficiently protective of public health. The science of and regulatory response to the climate change issue could overwhelm even these significant regulatory concerns; already, the global insurance industry has begun to plan as if the climate change risk were real. Living sustainably within environmental limits will require a substantial increase in energy efficiency and renewable energy sources.[19]

Current IRP and utility-planning approaches at least open the possibility of integration of these emerging environmental concerns with traditional cost-and risk-minimizing generation and other system-planning concerns.

NEES's recent NEESPLAN 4 commitments to DSM, renewables development, and sophisticated options-based wholesale power contracting[20], which seek to balance near-term system rate concerns with medium-to long-run environmental risks, stand as examples of how this integration can occur.

In a retail wheeling world, by contrast, the utility with the cheapest near-term price wins. This is hardly likely to encourage medium- to long-range development of power system end-use efficiency or development of low-environmental-impact capacity in anticipation of future environmental regulation when such investments result in a higher near-medium term price. It is instead likely to encourage the continued operation of

---

[19] *See* S. Brick, "Impending Regulatory Changes for Ozone, Sulfur Dioxide, and Air Toxics" (MSB Energy Associates, Nov. 1993); "Pollution, Pollution: Federal Air Standards Permit Dangerous Particulate Levels," *Scientific American*, Nov. 1993; D.W. Dockery et al, "An Association Between Air Pollution and Mortality in Six U.S. Cities," 329 New England J. of Medicine 1753 (Dec. 9, 1993); E. Londen, "Burned by Warming: Big Losses from Violent Storms Makes Insurance Industry Take Global Climate Change Seriously," *Time*, Mar. 14, 1994, at 79.

[20] *See* New England Electric Systems, NEESPLAN 4: Creating Options for More Competitive and More Sustainable Electric Service (Nov. 1994).

depreciated fossil fuel plants as long as possible. For new capacity, utilities and other producers will—as they have in the U.K.—prefer investments with low capital requirements and short lead times, such as natural gas-fired combustion turbines and/or combined-cycle generators. Retail wheeling will also—as we have noted earlier—likely pose severe challenges to continued operation of nuclear plants. Whether this is a "good" or a "bad" thing environmentally on its own terms, nuclear shut-downs could require significant construction of new capacity to fill the gap—which, under the pressures of a fragmented retail market, is likely to use fossil fuels only.

There are a number of problems raised by this scenario. First, while gas-fired plants are cleaner than other fossil plants, they most likely will not constitute by themselves an answer to significant environmental challenges; for example, on the basis of New England data it appears that even substantial gas repowering would not bring the region into compliance with the goal of 50 percent or more carbon emissions reduction believed by most climate scientists necessary to achieve climate stabilization. Second, the gas would expose customers to the risk of rapid increases in the price of natural gas as well as to environmental and cost risks associated with increased regulation of fossil fuel emissions. Third, there is the risk that a substantial amount of social capital will be wasted should the environmental picture in turn require substantial replacement of new "clean" coal and gas plants with renewables and efficiency—or, alternatively, that such a necessary transition will be resisted by powerful interests with a direct stake in preserving the economic viability of this "first wave" repowering.

The standard response to the concerns over the incompatibility of sustainable energy decisions with a short-term retail price-driven environment is that such considerations can be dealt with through external taxation schemes—either emissions taxes or T&D taxes to support DSM and renewables—or through direct emissions controls. But such an approach is unlikely to avoid either the problem of costly investment in generation made subsequently obsolete by successive waves of environmental regulation or taxes, or the practical problem that newly vested interests will successfully resist such regulation and taxes. In addition, taxation schemes are more likely to produce only token "set-asides" for renewables and DSM (as in the U.K.) than an environment that requires a rigorous cost-and risk-weighted utility-specific comparison of these resources to conventional generating options. Decoupled from power system economics, such investments will be firmly ghettoized as "social programs," as if they offer no internal power system benefits. This marginalization of environmentally cleaner options runs contrary to the worldwide corporate trend towards "industrial ecology," and related approaches.

It is indeed ironic that, just at the moment when cost-effective "pollution prevention" approaches have gained prominence in national, state and corporate policy, we would abandon perhaps the most powerful opportunity available to implement that approach—in the electric power sector.

Compatibility with Distributed Generation

The retail wheeling vision of the U.S. power industry's future imagines a world in which undifferentiated commodity bulk power from units sized and optimized to short- and medium-term contract flows are wheeled to the highest bidder. It is hard to reconcile this vision with the emerging opportunities inherent in distributed generation, which relies on strategically-placed modular generating technologies such as wind plants in remote load areas, fuel cells mounted in hotel basements, customer-site cogeneration, and rooftop PVs.[21] Aside from lowering the environmental burden of today's electric power systems by reducing emissions and local transmission—and generation-siting impacts—the transmission facility savings associated with such strategically distributed generation are likely to be substantial.[22]

Yet it is unlikely that utilities or other entities would make investments in such distributed resources in a retail wheeling environment, since their economics rest in large part on system-wide strategies to avoid transmission investment and upgrades and optimize the deployment of generation system-wide. In a balkanized world of retail wheeling, these investments might make little sense on a stand-alone basis. Unless elaborate market mechanisms can be developed to capture and integrate the system value of these distributed technologies, a large economic and environmental opportunity will be precluded.

---

[21] C. Flavin and N. Lenssen, "Reshaping the Power Industry," in L. Brown, et. al., *State of the World 1994* 72–75 (New York, 1994); D. Weinberg, D. Moscovitz, T. Austin, C. Harrington, and C. Weinbert, "Future Utility and Regulatory Structures: If You're Going Any Road Will Get You There," (Regulatory Assistance Project, Dec. 1993) at p. 4–7.

[22] Flavin and Lenssen, supra, note 20, at 75.

Chapter 7: The Political Economy of Retail Wheeling

Political Legitimacy

The very existence of this journal is testimony to the political salience of electric power system issues, and particularly their interface with environmental concerns. Public concern about and reaction to power system environmental impacts and generation—and transmission-siting—decisions have transformed the electric power scene in the last decade. Those concerns are likely to intensify rather than to abate. By wide margins, the American public continues to articulate a preference for exhausting low-impact renewable energy sources and energy efficiency before conventional generating technologies are deployed, even at a somewhat higher cost.[23]

The institution and political response to these environmental and consumer concerns has been the creation of public energy facility-siting processes and integrated resource planning. While there is rarely complete public satisfaction with the results of these reviews, there is broad acceptance of the process. Retail wheeling and broader industry restructuring proposals with a retail wheeling component locate generation and siting decisions in the realm of short- to medium-term retail markets, and therefore nullify public participation in the resource selection process. Under retail, there is no forum or criterion in or by which to justify a particular generating plant or transmission line as "least cost," or as the best of the long-run alternatives. There is only the aggregation of thousands of retail generation contract decisions. In a retail wheeling world, there is no "big picture" into which any incremental generation or transmission decision can be coherently explained or justified. A facility is "needed" because a developer believes he or she can make money on it.

Aside from presenting the legal problem of reconciling this balkanized decision-making approach with existing state siting statutes requiring balancing of environmental cost and "need" factors, retail wheeling thus presents a major political legitimacy problem. It remains to be seen whether the public will accept major siting decisions without the opportunity for meaningful public participation and discussion of facilities economics and need. Our experience suggests that, given the high level of public literacy and concern about energy issues, they will not.

---

[23] *See*, e.g., "America at the Cross Roads—A National Energy Strategy Poll," Vincent Bregelin Research/Strategy/Management, Inc. and Greenberg-Lake Analysis Group (Jan. 11, 1991).

At their core, retail wheeling and more ambitious U.K.-style restructuring schemes rest not only on short-term (and, we argue, mistaken) self-interest, but also on a particular political philosophy: the view that markets always "know best" and that the aggregation of private contract decisions is by definition democratically superior to planning of any sort. At a minimum, some retail wheeling advocates argue, efforts to influence the long-run shape and environmental impact of the electric power system should be confined to the legislative arena, and may not be internalized through economic regulation.

While it is difficult to dispute this Hayekian article of faith or any political philosophy directly, we note here simply that this radical market vision is quite inconsistent with the actual American consensus and experience. First, the notion that the outcome of the shifting and instantaneous "referendum" of the marketplace is by definition good public policy runs directly against key elements of American public life—not least the reflective filtering process of republican representation and the tradition of delegation to expert agencies to resolve detailed matters that legislatures are not prepared to decide. Second, a significant number of American states have given explicit legislative direction to utilities and their regulators to consider and balance long-term cost and environmental goals, while leaving the details to the commissions to work out. Third, despite rhetorical claims that integrated resource planning represents the "capture" of utility commissions by environmental "special interests," the move towards energy efficiency and greater environmental concern in the regulatory process in recent years is in fact, as noted above, quite broadly congruent with the views of the American public—hardly evidence of an "undemocratic" insider conspiracy. Finally, we find it contradictory that political economists who consistently bemoan the intrusion of environmental considerations into the economic regulatory process are rarely shy about suggesting that economic analysis should be a major focus of the environmental regulatory process.

In short, a pragmatic political economy—the notion that markets should be used as an efficient means but not always to determine important ends—rather than mechanistic ideology has been the hallmark of the evolution of the power system debate and regulatory practice. The political legitimacy of that practice will, we believe, be drawn into serious question under a retail contract-driven regime.

The Road from Here

As we noted at the outset, there is always a danger in fighting the last war. Designing an electric power regulatory system to address the problems of nuclear imprudence costs rather than the problems and opportunities that lie ahead would, in our view, be a tragedy. The choice is not between the pre-IRP and pre-wholesale competition regulatory system of the 1970s and retail wheeling. It is between retail wheeling and the emerging reality of a regulatory system that incorporates aggressive wholesale competition and forward-looking environmental risk mitigation to ensure that cost and risk are minimized. (Indeed, when such controls were in place, as we have noted, rate differentials about which retail wheeling advocates now complain are significantly smaller today.)

The outlines of that alternative system have begun to emerge and have been debated extensively in these pages. First, such a system would require or encourage cost-effective wholesale competition, an agenda that is far from universally accepted. Second, writedowns of utility plant or termination of existing IPP generation commitments should be addressed politically head-on, where the winners and losers can be openly identified and can debate, rather than through the indirect device of retail wheeling. Advocates wishing for the latter may well be disappointed if they get their wish.

Third, medium- to long-term environmental risks need to be reconciled more effectively with power system strategies to avoid consecutive wasteful waves of power system investment and subsequent obsolescence.

These recommendations surely lack the sex appeal of breathless trade press accounts of retail wheeling. But building peace is always less interesting than waging war, especially when it's the last one.

CHAPTER 8

Progressive Choice: The Customer as Regulator

Philip R. O'Connor, Ph.D.[1]
Managing Director
Palmer Bellevue/Coopers & Lybrand
Terrence L. Barnich
Craig M. Clausen

"Freedom's just another word for nothin' left to lose."[2]
Janis Joplin, 1971

"Freedom's just another word for somethin' more to choose."
Terry & Phil, 1993

The late lamented blues singer from Port Arthur, Texas, had it a bit wrong, as the recently liberated populations of Eastern Europe might attest. Freedom—the ability to exercise choice, personal, economic and political—may not be easy, but it is better than not having any choices at all. If any unifying theme characterizes human development in the past several centuries, it has been the struggle between the individual's drive for greater freedom and choice and the desire of some to deprive individuals of their freedom and choice on behalf of some vague notion of collective good. This basic struggle now envelops the utility industry.

---

[1] Terrence L. Barnich was appointed to the Illinois Commerce Commission in November, 1989, for a term ending in January, 1994, and served as chairman of the Commission from October, 1989, through February, 1992. Philip R. O'Connor has been chairman and CEO of Palmer Bellevue Corporation since January, 1986, and previously served as chairman of the Illinois Commerce Commission from January, 1983 to December, 1985. Craig M. Clausen served as executive assistant to the Chairman of the Illinois Commerce Commission from November, 1988, through February, 1992, and is currently senior policy advisor to Commissioner Barnich.

[2] From "Me & Bobby McGee" written by Kris Kristofferson, (New York, NY: Columbia Records, 1971).

The Electric Industry in Transition

In its day, the vertically integrated "natural" monopoly utility—including telephone, gas and electric—delivered new and wonderful choices to people. Consumers had new options available to replace the telegraph and the mails, coal and oil for furnaces, and town gas for lighting (which, in its turn, had offered an alternative to candles and whale oil lamps). The regulated utility monopoly endured over the years precisely because it offered choices and options in keeping with the drive toward greater freedom and the enhanced quality of life that freedom of choice brings.

The past two decades of developing competition in the telephone, natural gas and electric industries have been elaborately discussed elsewhere and require no further attention here.[3] Suffice it to say with respect to all three industries, the process is well underway for the disestablishment of the vertically integrated utility as the sole legitimate model for the delivery of these services. AT&T's divestiture of the Bell Telephone Companies; the passage of the Natural Gas Policy Act of 1978; the Federal Energy Regulatory Commission's Orders 436, 500 and 636; and enactment of the Energy Policy Act of 1992 all represent official recognition and confirmation of that basic process.

The argument within industry and regulatory circles is no longer whether there will be increased competition but how that increase should be managed, what role regulators should play and how transition costs can be smoothly and fairly apportioned. For regulators, the ultimate challenge will be to replace profit regulation with customer choice as the central theme of utility oversight and the measure by which regulatory action is judged.

---

[3] *See*, for instance, J. Dasovich, W. Meyer, and V. A. Coe. *California's Electric Services Industry: Perspectives on the Past, Strategies for the Future*, (San Francisco, CA: California Public Utilities Commission, Division of Strategic Planning, February, 1993). This report presents an outstanding history and analysis of developments in the electric industry the past two decades. The report also served as the centerpiece for a series of *en banc* seminars held by the California Commission to consider the range of possibilities for the reformulation of regulation and restructuring of the electric industry. Our chapter here is based on two separate papers presented, at the invitation of the Commission, as the keynote presentations for the first two seminars. On April 22, 1993, Mr. Barnich presented a paper entitled "Challenges and Opportunities: California's Electric Services Industry" and on May 25, 1993, Dr. O'Connor offered his paper, "Progressive Choice: A Model for Consumer Choice in the Electric Power Industry."

Premises for Change—Progressive Choice (PC)[4]

The underlying premises of the progressive choice model are simple. Evaluating any particular feature of a practical PC program is straightforward, assuming that faith is kept with the underlying theory. A key PC objective is to distill the essential elements from a regulatory system that has become mired in minutiae. Regulatory initiatives or utility proposals to accommodate change have been entangled in much of the same sort of hair splitting and juridical sophistry that has come to characterize the nation's court system.

The current regulatory framework is out of sync with competitive realities in the electric market.

♦ Competition in the wholesale generation sector has brought about the demise of the vertical monopoly as the sole legitimate model for the utility.[5] Other elements of the "natural" monopoly, such as transmission and distribution, may also prove vulnerable in light of technological change and global competitive economic forces that give customers increasing choices to leverage utility services.

♦ Conventional regulation has distorted the relationship between underlying costs and ultimate customer prices. The regulatory process, often with the acquiescence or connivance of utilities, has inflicted significant rigidities on pricing that conflict with the notion of flexibility in a competitive market. Distorted price signals flowing between consumers and utilities obstruct optimal efficiency in operation, investment, and consumption decisions.

---

[4] Progressive Choice (PC) attempts, among other things, not merely to usurp the use of the PC initials but to merge the values of the personal computer and political correctness. Progressive choice is fundamentally information-based, relying on individual customer decisions, driven by access to timely, accurate information (through relatives and descendants of the personal computer) to achieve politically correct results of lower prices, more environmentally conscious energy production and usage, and the exaltation of individual choice.

[5] A report by Bechtel Power Corporation, *Outlook for U.S. Power Markets*, (San Fransico, CA, July, 1993), indicates that an increasing percentage of baseload/cycling generation plants is already currently being built by non-utility generators (NUGs). Furthermore, Bechtel forecasts that NUGs will meet half the nation's generation capacity demands in the next ten years. This prognostication is not out of line with other forecasts in the industry.

- Inconsistent regulatory attitudes about the role of competition produce a hit and miss approach to the rules of entry and exit by electric market competitors. Barriers to both entry and exit inhibit the normal playing out of competitive pressures in the market.[6]

- Equity holders and investors of utilities, as well as those of new electric industry competitors, can increasingly find regulation a risk producer rather than a risk mitigator.[7] Market forces, the utility commission's role in planning, the fragility of deferred revenue recovery items, and mandates for particular resource acquisitions converge to produce a volatile mix of regulatory and market risk.

- The mismatch between the rules of the regulatory game and the realities of the market encourage many customers and existing utilities to develop regulatory exit strategies.[8,9] Like people going over the Berlin Wall, customers and utilities seek the freedoms they have heard exist on the other side. Regulators should not accept employment as border guards. Instead, their job should be to make the system attractive enough that customers will not seek to escape.

---

[6] Alfred E. Kahn, "Telecommunications, Competitiveness and Economic Development — What Makes Us Competitive?" *Public Utilities Fortnightly*, September 13, 1990.

[7] Philip R. O'Connor, "Utility Regulation in Illinois: Uncertainty as a Regulatory Product." *Twenty Years of Energy Policy: Looking Forward Toward the Twenty-First Century (Proceedings of the Twentieth Annual Illinois Energy Conference)*, (Chicago, IL: The Energy Resources Center, University of Illinois at Chicago, 1992).

[8] *See*, Edward J. Tirello, Jr., "Traveling Light," *Public Utilities Fortnightly*, July 1, 1993.

[9] As the electric industry has matured and competition has intensified, margins have narrowed and authorized and actual returns on equity (ROEs) have declined. Domestic energy companies have responded by seeking to diversify in order to seek out profitable un-regulated ventures abroad. Rigidities in pricing and other terms of service, all controlled by regulators through elaborate proceedings encourage both buyers and sellers to seek contexts in which greater freedom and flexibility can be exercised. As larger customers seek to escape the imposition of cross subsidy responsibilities, their ability to relocate, shift production or to self-generate all present competitive pressures on local utilities to offer concessions of various types.

Chapter 8: Progressive Choice

Global forces are reshaping the electric business just as they have reshaped other industries, regulated and unregulated. The list of developments driving basic changes in the world economy is long and varied. Running through any such list, however, is the theme of choice and the customization of services through the use of amazing advances in the creation, communication and manipulation of information—all at the fingertips of hundreds of millions of ordinary people and businesses.[10] The collapse of the destructive communist experiment has spread both "American" culture and the market ideal far and wide. Now, as it become the world's "fuel of choice", electricity will be seen in the same terms as other services—one that must be provided competitively and tailored to meet individual customer needs.[11] The globalization of finance and the relatively uninhibited flow of money around the planet mean that electric industry financing needs and opportunities must be on a worldwide competitive footing.[12]

The utility regulatory system requires a fundamental overhaul because mere reform cannot lead to the full blossoming of competition. Only genuine competition, rather than simulations of competition, can actually deliver the fullest measure of consumer benefits. Utility and regulator forecasting techniques cannot possibly predict the combination of efficiencies and innovations that will emerge from a competitive environment. Government regulators, especially, lack complete information about the markets they are regulating. Regulatory and utility planning models tend to be linear and relatively undynamic, unable to accommodate the entrepreneurial response to the complex of risks and rewards that are characteristic of more fully competitive industries. Regulatory proceedings seeking proof of the precise consequences of a movement toward competition are a futile exercise, asking questions that can be

---

[10] B. Joseph Pine,II., *Mass Customization: The New Frontier in Business Competition*, (Boston, MA: Harvard University Press, 1993).

[11] Peter F. Drucker, *Managing For The Future: The 1990s and Beyond*, (New York, NY: Truman Talley Books, 1992).

[12] Philip R. O'Connor and Wayne P. Olson. "Global Challenges in Energy and the Environment," (Chicago, IL: Palmer Bellevue Corp., October, 1992).

answered only by time and experience.[13] The posing of such questions will often be less a sincere effort at inquiry than an effort to constrain competition within narrow bounds. Use of the impossible-to-answer questions is the regulatory ploy of those seeking to limit action to incremental reform around the edges rather than carrying out fundamental (and sometimes radical) changes. The fundamental change we envision, which forsakes the central role of profit regulation and relies instead on competition, is justified by the same faith in competition on which we operate elsewhere in the economy.

A plan for fundamental change should take into account the nearly immutable patterns of change in regulated industries as they move toward competition.[14]

◆ Incremental change. Fundamental change in the electric industry need not be made with blood and iron. Change in regulated industries has proven to be incremental. The system does not change overnight, but bit by bit. However, market forces cause it to do so at ever-accelerating speeds, leading to a new model, which then itself achieves a certain stability. Regulatory bodies are unequipped with gear shifts and are unable (or unwilling) to accommodate their pace to the market driven change and thus, all too often, end up acting as drags on the procession.

◆ Entropic change. Entropy begins to characterize the system with the entry of brand new players who take slices of the market, eschewing the costly and impossible effort to replicate the entire range of services provided by the existing utility. Ironically, among the most important new players are the few existing utilities who decide quickly that change is coming and alter their business strategies in order to

---

[13] Charles Stalon, "Decision Making, Information Overload and the Pursuit of Legitimacy," *Proceedings of the 103rd Annual Convention of the NARUC*, (Washington, D.C.: National Association of Regulatory Utility Commissioners, 1992).

[14] *See*, Philip R. O'Connor and Gerald M. Keenan. "The Politics and Policy of Access to the Electric Utility Transmission System," *Public Utilities Fortnightly*, July 7, 1988, for a more complete discussion of these four themes in the transition of the electric industry.

meet the competitive challenge. These incumbents also create the case for overruling the objections of more recalcitrant incumbents.

◆ Moving prices to cost. The new players force prices to move toward cost. The system of cross subsidies, which characterizes regulated monopoly, becomes unsustainable. Importantly, the subsidies themselves help unravel the closed market because once the new entrants' products and services become available the customers paying the subsidies actively look for alternatives where prices do not include subsidies.

◆ Resistance to change. Finally, many utilities and regulators resist change in an ultimately failing cause. In the process, the regulators tend to reinforce the subsidy system that increasingly disadvantages the regulated firms by providing advantages to their new competitors, who are more free to meet customer needs. Resistance not merely delays customer choice, it is costly for the incumbent utility and inhibits its ability to offer choice even when the regime of choice achieves hegemony. For the regulator, resistance to change ends up disrupting the realization of the very social or political goals for which the subsidies were originally created as the incumbent utility is increasingly unable to respond flexibly, if at all.

The Principles of Progressive Choice

Just as there are four premises for offering PC as the mechanism for basic change, there are four principles on which a regulatory format of PC in the electric business is based. All four focus on regulator-led enhancement of choices to provide "protection" for consumers or investors, not regulator-based denial of choices.[15]

Percolate the benefits of competition. Recognizing that competitive forces are not equally distributed throughout the electric market, a PC program should focus on percolating or flowing through competitive forces from competitive market segments to segments in which competition

---

[15] *See*, Philip R. O'Connor, "The Protection of Core Customers: Enhancing Customer Choice," Presentation at the Annual Convention of the National Association of Utility Regulatory Commissioners (Los Angeles, CA, November, 1992), for a thorough discussion of the problem of "protection" as a counterproductive mindset for utility regulation.

is limited. This principle recognizes that many customers, especially larger ones, have the ability to extract concessions from the local utility on a variety of options ranging from DSM and process changes to relocation and the installation of self-generation. This is not a reason to forestall competitive inroads, but rather a reason to encourage their reach to a wider customer base.

Auto-pilot the change. Regulatory micro-management should give way over time to competition. The process of change itself should be designed to operate nearly automatically, with minimal regulator involvement in the "progress" of the movement toward competition once the process has started.

Facilitate the movement to a competitive market. In order to mitigate utility resistance, as well as to address equity issues in the transition to PC, regulators need to focus attention on important transitional issues that may have significant financial implications for existing utilities, such as depreciation rates for sunk investment.[16]

Rely upon new information technologies. Communication and information technologies, rapid advances in which have been expansively cultivated by several utilities, lie at the heart of the ability of customers and electric service providers to exercise choice and to tailor services for specialized needs. PC assumes that reasonably priced information technology will be available to permit most customers to receive and act on real time pricing information.

The Grand Caveat
PC has a "non-principle" in that it does not depend upon retail wheeling as an essential feature.

---

[16] With the advent of the technologically driven competitive environment in the telecommunications industry, the regulated companies were, in part because of uneconomic depreciation policies, carrying on their books assets that were grossly overvalued. Beginning in 1980 the FCC responded by altering depreciation methods for the regulated companies to more accurately reflect realistic and timely capital recovery. This included accelerated depreciation for "inside wiring," allowing companies to "expense" all "inside wiring" in the year it was incurred. *Uniform System of Accounts*, 85 F.C.C. at 818.

First, there is every reason to implement the choice standard without the delay inherent in a bitter and insufferably consuming debate over retail wheeling and the way in which it would be regulated ("refereed").

Second, dispensing with retail wheeling for purposes of the PC discussion does not mean the issue goes away. Some will continue to believe retail wheeling is just over the horizon while others believe it is some considerable distance away. No doubt the debate will proceed. That debate will either accelerate or become quieter as average costs and long run marginal costs begin to converge in more places around the country.

Third, recognizing that the retail wheeling debate will proceed, there is a certain discipline to be imposed on the rest of the discussion of choice by undertaking it without it being held hostage by the retail wheeling issue.

Finally, disconnecting the retail wheeling debate from the broader question may make it easier to consider a wider array of multiple structural models for the industry.[17]

The Operating Features of Progressive Choice

PC does not seek to replace one rigid system with another rigid one: PC is intended to be malleable and subject to change.[18] As time goes on, regulators can refine the basic design to assure continued and improved access to choices, given that there is likely to be a continued mix of monopoly and competition. However, PC would, at the outset, embody eight overarching principles.

---

[17] *See*, for instance, Ashley C. Brown, and Terrence L. Barnich. "Transmission and Ratebase: A Match Not Made in Heaven," *Public Utilities Fortnightly*, June 1, 1991, for a discussion of the potential for open access of transmission that would result from the "de-ratebasing" of transmission assets. One obstacle to retail and wholesale wheeling may be that transmission pricing has not yet matured to accommodate a competitive environment, perhaps due to the inclusion of transmission assets as an undifferentiated item in retail utility rate bases.

[18] "Progressive choicers" can take a lesson from Karl Marx here. His notion of "praxis" provides for activist theory rooted in the principle that theory must be constantly revised by experience and practice. To the extent that the theory seems a bit "off" from reality then the theory may need amendment to address new-found facts.

An "Osmotic" Core/Non-core Market Segmentation. PC would cure the most significant flaw in efforts to establish a bright line between "core" and "non-core" markets. It has not been clear whether the distinction between customer groupings has been based on "protecting" those customers who have limited choices, or on limiting the share of the customer base that actually has access to choices. Certainly, many so-called "core" natural gas and telecommunications customers could find better prices and services in the market than they currently receive under the protection of regulation, if allowed to do so. Increasing their right to exercise choices will probably attract more competitors into the market to serve them and cause their utility to seek the flexibility necessary to satisfy these customers.

Under PC, those who start out in the core category are able, of their own volition, to move to the non-core category without having to seek regulator or utility approval. This process might be thought of as an "osmotic" flow of customers from the core to the non-core market through a "permeable membrane", a membrane purely economic in nature rather than regulatory. The ability to move through the membrane to non-core status would be a function of an individual customer's own belief in his ability to function in the non-core, more competitive arena. Available technology and a customer's evaluation of the economics of moving would govern these choices, as well as the willingness to forego the traditional protections of "core" status.

This "permeable membrane" could include some regulatory barriers to moving back to core status, such as re-entry fees. But core customers would no longer be hostages, prohibited by law or rule from changing their status.[19]

The PC core/non-core distinction would not center solely on the characteristics of the consumer (such as residential versus industrial). Rather, as in the classification of some telecommunications services, the PC

---

[19] This sequestering of customers by characteristics has been somewhat provocatively called "regulatory apartheid" by Peter Huber, *et al.* (P. Huber, M. Kellogg, J. Thorne. *The Geodesic Network II: 1993 Report on Competition in the Telephone Industry*, (Washington, D.C.: The Geodesic Co., 1993), p. 1.27).

core/non-core (competitive/non-competitive) distinction would be shown in a matrix of at least two dimensions—of customers and services.[20]

Unbundled Services at Negotiated Prices. When some customers have increasing ability to extract concessions from the local utility, the relationship between the non-core customer and the utility should become characterized by flexibility and negotiation. Non-core customers and all energy service providers (including the utilities) should be permitted to negotiate the prices and terms for all services. This negotiating process will lead to the unbundling of current services and the offering of new ones. Services that start out as bundled and core can be unbundled and offered competitively to some customers while continuing to be offered as bundled core services to others.

In a competitive market, the idea of unfairly discriminatory prices should be considered only in the narrowest of terms. Mere differences in price among similarly situated customers, even large differences, should not form the basis for intervention. Only differences based on invidious

---

[20] Innovations in classifying services as competitive (and thus, non-core) irrespective of the class of the customer using the service, have been employed for some time in the telecommunications field. Most conspicuous has been cellular telephone, which many jurisdictions simply have de-regulated. PBX and Centrex services for small business customers are largely de-regulated and Integrated Services Digital Network (ISDN), to the extent it is deployable, will follow suit. In mid-1993, Indiana Bell Telephone filed an alternative regulation plan called "Opportunity Indiana" which allocates services to various categories, competitive, discretionary and basic. In addition, rates for basic services for residential customers would be guaranteed to rise at a rate somewhat lower than general inflation.

The movement toward full competition within the local telephone exchange is well underway. The first articulation of a state regulatory strategy to accommodate this movement appears in Terrence L. Barnich, Craig M. Clausen, and Calvin S. Monson. *Telecommunications Free Trade Zones: Crafting a Model for Local Exchange Competition*, (Springfield, IL: Illinois Commerce Commission, January, 1992). On August 3, 1993, the Federal Communications Commission moved another step closer to completely opening the access portion of the local market through its switches access interconnection order (FCC Order 91–141).

and genuinely unacceptable reasons (such as those addressed in civil rights laws) should make government intervene.[21]

Under PC, the initial group of non-core customers will be identified by their ability to negotiate with the utility. Size, demand characteristics or traditional class grouping will not necessarily be the defining elements. More important will be the characteristics that would indicate that choice already is or soon could be an important part of the relationship between the utility and the customer. Customers who can easily self-generate, switch fuels, and shift production to other locations or potential customers who can decline to come into the market area all have bargaining power. There will be a premium on creating an initial group of customers, which would begin to define a different relationship between the utility and its customers. This initial process of classification can be made complicated if regulators, utilities and intervenors choose to make it so by requiring enormous amounts of analysis to determine the membership of the first non-core group. The criteria should be simple and straightforward. In addition, the first couple of years of membership in the non-core group could be allowed on a trial basis as a way of encouraging customers to volunteer for non-core status at the beginning of PC.

---

[21] *See*, Phillip Areeda and Donald Turner, "Predatory Pricing and Related Practices Under Section 2 of the Sherman Act," *Harvard Law Review*. Volume 88: pp. 697–733 (1975). Predatory pricing, wherein a supplier sets prices below his actual cost in an effort to run his competitors out of business and recoup his losses by imposing monopoly prices later on, is a basis for anti-trust law and a traditional touchstone for unfair trade practice allegations. However, this concern results in turning a blind eye toward the welfare benefits that flow from competition's rigorous discipline to move prices downward. Confusing price and policy change in a competitive market for monopoly behavior is a prescription for higher prices and lower service quality over the long run. *See*, Robert H. Bork, *The Antitrust Paradox: A Policy At War With Itself*, (New York, NY: Basic Books, 1978).

Alfred Kahn once commented that "[o]ne of the most damning condemnations of motor carrier regulation was the demonstration that of all the motor carrier pricing decisions made by the ICC during one year, 95% involved complaints that the prices were set too low and only 5% the setting of ceilings." Thus Kahn shows empirically that traditional "protect the consumer" regulation sets the whole public policy basis of competition on its head.

Chapter 8: Progressive Choice

Real-time pricing. Currently, most customers have little sense that electricity costs more or less to make depending on when it is produced. Regulated prices obscure this fact. Even larger industrial customers on time-of-use tariffs tend to see only gross price/cost relationships in their billings. Average cost pricing just does not move pricing signals back and forth between customers and the utility quickly or accurately enough for the needs of a modern, competitive and information-oriented market place. This lack of information is a key cause of the poor load factors that characterize many utility systems. Poor pricing undercuts the implementation of cost-effective, sustainable DSM. One effective way to begin breaking through this barrier to customer choice is to make real-time pricing available across all customer groupings—at the election of the customer.

Micro-electronics and inexpensive methods of communication allow customers to control their usage and to shift load, automatically or with little direct action, in reaction to different price levels. It is by no means essential that customers have absolutely precise production cost information as long as they know the price they will be charged at any specific time. A pyramided price scheme with seven differentiated levels for each day would be a dramatic improvement over the single price now applied to most customers or the two or three prices larger customers now see in the course of a 24-hour period.[22] If electric utilities are reluctant to make the information infrastructure available for real-time pricing, local telephone and cable TV companies, as well as Radio Shack, will probably be interested in doing so.[23]

---

[22] Already, important real-time pricing experiments have been conducted or are ongoing around the country, including ones at Niagara Mohawk, Georgia Power, Consolidated Edison, Entergy Corp., American Electric Power (AEP), Pacific Gas & Electric and at Southern California Edison. All of these real time pricing programs have been strictly experimental, limited in scope, directed at a variety of customer groupings (including residential in the case of Entergy and AEP) and have produced mixed results. Taken together, however, they are harbingers of things to come. Technology is rapidly moving to allow customers to inexpensively acquire current pricing information, alter consumption and to have that change measured in real time.

[23] Even now projects such as the venture with First Pacific Network and Entergy are under way to bring fiber optic electronics to the home using the electric utility rights of way to provide "smart" energy management for Entergy's residential customers.

Real-time pricing can also be used to move core customers into the use of non-core services and eventual passage into complete non-core status. Core customers, whose prices may be otherwise set under a tandem pricing plan (described below) could opt for real-time pricing as a non-core service.

Real-time pricing could have major environmental benefits, under the sorts of smog precursor (emissions) trading schemes that have been developed in Southern California by the South Coast Air Quality Management District (SCAQMD) and in the Chicago area by the Illinois Environmental Protection Agency (IEPA). The internalization of environmental costs through specific trading mechanisms might well be reflected in real-time pricing of electric power because smog-related emissions' effects on the local environment are time specific. Movement to an aggressive real-time pricing program can be accompanied by aggressive marketing of associated electronics and DSM services carried out under a PC regime of competitive, unregulated prices.

Tandem Pricing.[24] Price cap regulation is intended as a way of weaning traditional regulation from its obsession with profit and returning regulation to its primary purpose, eliciting fair prices for desired service.[25]

---

[24] Tandem pricing is partially inspired by the concept of "leveraged pricing" first developed by Dr. J. Cale Case of Palmer Bellevue Corporation. Leveraged pricing was first advocated for application to pricing of core telephone services for residential customers, linking the pricing of individual residential services to unbundled service offering prices in the competitive, non-core market, where the services have similar underlying costs. Tandem pricing, while perfectly capable of linking prices on a service-to-service basis, is more centered on accommodating the linking of baskets of services with one another, accepting some greater disparity in the underlying cost relationships. *See*, J. Cale Case. "Leveraged Pricing: A Better Alternative For Telecommunication Regulation," *Proceedings of the Sixth NARUC Biennial Regulatory Information Conference. Volume III: Telecommunications, Water and Transportation Papers*, ed. David W. Wirick, (Columbus, OH: The National Regulatory Research Institute, 1988).

[25] Thomas K. McCraw in his book, *Prophets of Regulation*, (Cambridge, MA: The Belknap Press of Harvard University Press, 1984) shows that utility regulation did not have its roots in complicated profit regulation. It was intended to focus on assuring the maintenance of prices which were fair in relationship to the service being provided. The departure from this standard and the gradual adoption of rate-of-return regulation predicated on the regulation of profit on investment in property grew up in response to court decisions around the turn of the century. Utility regulation has often been the effort to reconcile financial and economic theory and practice with notions generated in the courthouse.

Price caps, based on initially just and reasonable rate levels, are set and then indexed to an inflation measure, less a defined productivity factor, thus creating incentives for cost control.

While price caps represent a significant advance over rate-of-return regulation, they still represent an effort to simulate market forces rather than injecting a significant dose of direct market medicine.[26] A more direct way would be to adopt tandem pricing in which price caps would be based upon an index measuring prices in the non-core market. Afterward, price change developments in the competitive market would move prices in the core markets as well. Rather than requiring regulators to choose an appropriate productivity factor as a discount to the inflation index, tandem pricing lets the competitive market select the productivity factor. Tandem pricing can assure that core customers get the benefits of competition, with little utility or regulatory dilution.

In the simplest model of tandem pricing, prices would be initially set for core services provided to core customers on a just and reasonable basis. At the same time, a benchmark price level for the non-core market would be ascertained. Thereafter, core and non-core prices would move in correlation with one another. Prices need not to move up or down on a one-for-one basis. Prices in the core market could move at one-half or three-fourths the rate of changes in the overall basket of non-core prices, because the full measure of competition is not likely to exist at each level of the market. In addition, keeping full flow-through of non-core market price changes from core customers will encourage those customers on the cusp between core and non-core markets to choose to move into the non-core market. Tandem pricing should almost function like an auto-pilot pricing mechanism in the transition from regulated to competitive pricing as technology and the market require.

---

[26] *See*, Terrence L. Barnich, "The Challenge for Incentive Regulation," *Public Utilities Fortnightly*, June 15, 1992.

The devolution of rate-base assets to competitive status and the role of affiliate transactions.[27] To the extent that utilities and customers are given the freedom to exercise the choice to convert to market-based rather than regulated relationships, it is also reasonable to allow a utility sufficient freedom to organize its assets (including the capital structure associated with those assets) so as to conform to competitive market behavior better. Form should follow function.

The Energy Policy Act of 1992 wisely left to the states the authority to determine whether a currently rate-based generation asset could be devolved to electric wholesale generator (EWG) status. With prior approval, a utility can transfer ownership of a generation unit (presumably for consideration under state commission oversight) to an affiliate EWG company. While the use of tandem and real-time pricing might diminish the interest of a utility in devolving generation, the option should be clearly articulated and the regulator's openmindedness expressed in the development of a PC program. Core customers are not disadvantaged by devolution under PC because the utility would be obligated to move core prices in tandem with non-core prices.

The corollary to permitting devolution of rate-based assets and to the general principle of encouraging flexible response by utilities is an openmindedness toward affiliating EWG transactions between utilities and their affiliate EWG companies. One important aspect of such affiliate transactions may well prove to be more comprehensive than customary energy service deals between utilities and their industrial and commercial customers. For instance, under PC, regulators ought to be open to deals in which industrials serve as hosts and partners for cogeneration or other power plant developments by utilities.

---

[27] The use of the word "devolution" for the movement of utility rate based generation assets (first appearing in Philip R. O'Connor, *Competition in the Electric Utility Industry: Sunset Series Monograph #15*, (Springfield, IL: Illinois Commerce Commission, 1985) is intended to provide a looser concept than that implied by use of the word divestiture under which the generation assets would move out of the corporate family of the utility into independent ownership. The 1992 Energy Policy Act amendments to PUHCA explicitly provide for devolution, regrettably without use of the word, by which utility rate based generation can be spun off to Electric Wholesale Generator (EWG) status, as long as state regulators agree.

Chapter 8: Progressive Choice

Competitive selection of supply and demand-side resources. One area of conflict that is bound to become sharper if there is a movement toward PC is the role of regulators in IRP and mandated resource acquisition and demand-side programs. If the government mandates particular investments and expenditures, the costs of many such mandates are likely to exceed those that would have resulted from competitively driven choices.

PC's underlying philosophy is that market forces result in better decisions than government intervention. But there may well be important goals that can better be achieved by a combination of government goal setting and/or prescription of method. PC allows for such intervention and prescription, but with the caveat that there is no free lunch; that price distortions should be kept to a minimum and social responsibilities be spread as widely across the market as possible.

PC ought to offer a better future for DSM than the current mode of regulation. Regulatory pricing would no longer shield customers from the actual costs of particular patterns of demand and consumption—including the real-time costs of environmental impacts. Conventional rate-of-return pricing has the perverse effect of pricing power too low when there is not enough and pricing it too high when there is too much. Various states have gone so far as to ban the inclusion of construction work in progress (CWIP) from rate base, thus exacerbating the situation. More accurate pricing through competitive and market forces would move DSM investment into the portions of the load curve that are in need of capacity. To the extent that kilowatt hour conservation is most valued by the market, then that is where resources should go. Similarly, to the extent capacity is needed, load shifting and peak shaving strategies can be implemented in a cost-effective way.

PRISM contracts (price-responsive industrial marketing contracts).[28] PC could employ a subset of competitive market resource acquisitions that would give the local utility the opportunity to exercise its classic value-added role of aggregating and blending different resources to serve disparate customer power needs in an environment characterized

---

[28] Depending on how this particular idea develops it seems possible that other meanings for the PRISM acronym might be developed which convey slightly different notions, such as Portfolio Reflective Incentive Sales Marketing contracts.

by competition. A Price Responsive Industrial Marketing Contract (PRISM) would be defined as one in which the utility acquires a specific set of power resources that can be packaged, blended, and priced to match an identified set of customer demands. Through PRISM contracts, utilities could acquire new resources through contracts with EWGs or other third parties, but the sales contracts would be balanced with a specific portfolio of complementary contracts on the customer side. Individual customers could avoid many of the risks and complexities involved in arranging bilateral purchases from other utilities or from PURPA-QF units.[29] Instead, the utility would play its traditional "blending" role.

The genius of the founding giants of the American utility business was not to be found solely in their scale-up of power plants and their financial creativity (such as the open-ended mortgage bond).[30] It was also in their recognition that a variety of power plant types could be brought together and coordinated to serve ever changing customer needs that would, over time, become fairly predictable in their aggregate pattern of demand. One of the critical complications today in moving toward a competitive market in electricity is that our conventional model of the vertically-integrated local monopoly electric utility compensated the electric company for its integrative (blending) role through the grant of monopoly status and subsumed the financial benefit into the return on the hard dollar investment rather than into a service fee that reflected a

---

[29] The Energy Policy Act of 1992 amendments to PUHCA, which created the category of Electric Wholesale Generators (EWGs), prohibits an EWG from engaging in retail sales. Therefore, this entire class of power producers would not be directly available to customers but would have to have their supplies mediated through a utility, either the one that served the area in which an industrial customer was located or one which somehow arranged for retail wheeling through a local utility to a specific customer. The PRISM contract, because it relies on the value-added blending role of the utility, avoids transgressing the Act's prohibitions on sham EWG retail transactions.

[30] Sam Insull, the unfairly maligned genius who, as much as anyone, built the modern electric utility, created new financial mechanisms that allowed the rapidly developing technology and its associated efficiencies to be reconciled with the need to raise large sums of capital for the development and deployment of a vast network of electric power infrastructure. *See* Forrest McDonald, *Insull*, (University of Chicago Press, Chicago IL. 1962).

profit.[31] Just as profit centers have moved from hardware to software and service in many other industries, so too should that transition be made in the electric business.

PRISM contracts are an outgrowth of a simpler idea in which utilities could be expected to engage in "accountant's wheeling," reflecting "mirrored" contracts with third party suppliers and with large individual customers. Mirrored contracts would be simple in the sense that a mirror is just reflecting an image. PRISM contracts would contain far more complex elements, including commodity electricity and a variety of other services.

The major difference between PRISM contracts and the old-fashioned utility role of meeting many differing needs with bulk resources will be twofold. First, the PRISM contract will not rely on franchise rights to underwrite the purchase but instead on customer contracts and merchant relationships that underpin the purchase commitment. Second, PRISM resources would be acquired with much lighter oversight, if any, by state regulators. *Post hoc* and useful prudence reviews would be unnecessary since the resource acquisition would be undertaken pursuant to corollary customer contracts. The newer IRP processes in the states could be revised to accommodate PRISMs merely by taking the implications of PRISMs into account when resources for the non-PRISM customer base are considered.

Management of transition costs—acelerated depreciation. Conventional rate making and the related accounting techniques may be increasingly unsuited to current industry conditions. Assumptions about monopoly have led to overestimates of useful economic life and excessive optimism about the ability of a utility to recover deferred revenues. The

---

[31] Participating in the May 25, 1993, panel discussion at the California Public Utilities Commission in San Fransico, Jeanine Hull, Vice President and Counsel of LG&E Power Systems, Inc. of LG&E Energy Corp., the independent power affiliate of Louisville Gas & Electric, opined that the idea of a profit margin or mark-up for the local utility on purchased power represented an unnecessary tax on EWGs. While perhaps true in a basic sense, the labeling of that mark-up as a tax does not solve the problem inherent in the loss of profit potential for local utilities in the provision of commodity electricity to distribution customers. That loss deters some utilities from seeking purchased supplies in lieu of owned generation. However, providing a profit opportunity by specifically pricing the "blending" task is one way of reconciling customer and utility interests.

willingness of regulators to address the more arcane areas of rate making accounting may have important implications for the ease or difficulty with which we cope with the order of the new world.

Some of the most important transitional problems in regulated industries involve accounting practices or conventions unsuited to competitive markets. These artifacts of regulation trap assets into vintage valuations that have little or nothing to do with their economic value. Regulated enterprises, during periods of transition are often expected to continue the commitment of "undervalued" assets to customers at vintage prices while being free to price only their most expensive assets at market rates.

By identifying particular parts of the asset base for which some sunk costs could be recovered on an accelerated basis, regulators could significantly reduce resistance by existing utilities to important competitive changes—perhaps even retail wheeling. The precedents are there and demonstrate that the process can work. For instance, at the federal level and in several states, telephone inside wiring was depreciated off the books within just a few years, eliminating a whole category of utility investment from regulatory attention and treatment.

Southern California Edison's (SCE) suggestion that its interest in the San Onofre and Palo Verde Nuclear Generating Stations receive accelerated depreciation treatment was directed toward this problem.[32] The plants have come to be viewed as assets whose useful economic lives may be shorter than their accounting life for rate making purposes. Whether the reasons involve the contemplation of large investments to keep the plant in good operating order or excessive increases in operating costs due to federal nuclear regulatory mandates, the company has a different view today of the likely future for the plant than that which it and regulators once had. SCE, like other electric utilities, is seeking the financial flexibility to meet the more competitive environment that almost everyone sees coming.

---

[32] In early 1993, Southern California Edison Company petitioned the California Public Utilities Commission to permit an "additional capital recovery" of about $75 million annually for SCE's interests in the San Onofre and Palo Verde nuclear stations. Central to the rationale for the accelerated recovery is that new generation technologies will soon be coming on line at $500–800 per installed kilowatt while these nuclear stations have embedded capital costs of $1,350 and $1,900 per kilowatt respectively.

Progressive Choice and the Pace of Change

Many may subscribe to the notion that the utility industries are subject primarily to long, slow change rather than abrupt change. While change is indeed incremental, it can nevertheless come quickly in relation to the industry's expectations. The past decade alone demonstrates how dramatic change can be in utility industries. The telecommunications business is fundamentally different today from just ten years ago, prior to the divestiture of the Bell Telephone companies. Numerous competitors are entering the local exchange market, as they have the long distance business. There is a real question as to where the action will actually be in telecommunications. Will customers soon control the network through sophisticated end-use equipment and software, dipping into a global network of networks to extract the desired information and signals?[33] Or will the network be the manipulator of information in addition to delivering information to right place?

Ten years ago, virtually every molecule of natural gas that moved in the interstate market was owned by the interstate pipeline transporting the gas and sold at prices regulated by FERC. Today, pipelines do not own gas at all, although their marketing affiliates may, and no gas now sold in interstate commerce is subject to federal economic regulation. Other important aspects of the gas market have changed dramatically as well, including the use of storage, the role of end-use customers and marketers arranging gas supplies in competition with local gas distribution companies.

In the electric industry ten years ago there was a huge backlog of utility-built and owned generating capacity that was driving rates up far faster than the overall rate of inflation. Independent power came in small increments and accounted for just a small percentage of the capacity that was in development. Today more than half of all new generating capacity is independent, and the law has been changed to allow independent developers and utilities to compete on a level playing field in the generation development market. The generation industry is now largely wide open to competition.

---

[33] George Gilder has suggested that the future of telecommunications will be characterized by huge capacity "dark fiber optic" pipes into which vast amounts of information will flow to their intended recipients pursuant to the commands originating within the end users' own computer-like equipment. See George Gilder, "Into the Fibersphere," *Forbes ASAP*, January, 1993.

The Electric Industry in Transition

If all of this in just ten years is not enough change, then perhaps only such years as 1492 and 1945 would satisfy such aficionados of paradigm shifts.

Implementing PC does not necessarily require a single dramatic regulatory decision. Pieces of PC can be undertaken on an individual basis over as little as two years if the process begins now. In that space of time customer choice could largely replace regulatory and monopoly dictates electric industry structure, prices, products and customer service.

CHAPTER 9

The Electric Industry in Transition

P. Chrisman Iribe
Executive Vice President
U.S. Generating Company

When considering the implications in the electric industry's transition, the question we all face is: as the electricity industry begins this upheaval, can past competitive strategies provide guidance for the future? If so, can they ensure that the benefits to customers do, indeed, outweigh the costs to those who have to negotiate the transition?

I believe they can. Therefore, today, rather than focus on whether there will be an aggressive, competitive generation sector of the electricity business here in the United States, I will focus on what we believe that competition will look like and how success will be achieved.

With these general questions in mind, let's look at how the generating marketplace has evolved, and let's review—in three stages—the lessons we've learned over the years.

- First, the independent power industry today is an established, mature industry worth studying for lessons learned. In maturing, IPPs—like other generators—have learned the key lesson of a competitive market: how to add value. This is especially significant now in light of the inevitable transition costs for stranded assets and for the organizational, industrial and human restructuring that we will all have to bear.

- Second, adding value means managing risks for customers, providing customers desired flexibility and by organizing our companies to succeed.

- Third, changes in the generation marketplace are accelerating the shift from long-term to short-term contracts that entail new kinds of risks. What this means to us generators is that we will have to simultaneously tackle new kinds of risks and find new ways to add value.

The IPP Industry Today

First, then, let's look at where the IPP industry is today. The industry's growth and maturity are demonstrated by the increasing size of its projects. Over the last two years, the proportion of new plants coming on line with capacities larger than 100 megawatts has increased by 25 percent. Other examples of this growth include:

- Industry leaders, such as my company, U.S. Generating (USGen) and AES Corp., have recently completed construction of coal projects with capacities in excess of 200 MW.

- Leading developers of natural gas facilities, like J. Makowski and others, now regularly construct gas plants with capacities in excess of 250 MW.

- Sithe Energies, of course, is completing construction of its 1,000 MW gas-fired Independence Project here in New York.

Increasing project size has also resulted in parallel increases in the size and complexity of project financing. In the late 1980s, a medium-sized project cost in the $200 million range, but now we have larger projects that cost $800 million and more.

This obviously increases the complexity of financing. Along with this, in the early 1990s the industry had to overcome the reduced availability of bank project financing. This was done by winning the confidence of public debt investors and rating agencies, which gave us our access to other capital markets today.

Finally, we have seen the emergence of IPPs that are structured, capitalized, and managed to bring three or more projects into construction every year. In USGen's case, very rapid growth since our start five years ago has led to our current $3.7 billion portfolio of projects in construction and operation. Of that $3.7 billion, $2 billion is in construction alone. Big projects, creative financing and larger portfolios of real assets are all signs of an industry that is growing and maturing.

Adding Value by Managing Risks

The IPP industry leaders over the last few years have grown by being successful in a very competitive marketplace. Generation companies succeeded by ensuring that our products added value to our customers. Now let's discuss the ways generators can add value.

Chapter 9: The Electric Industry in Transition

And when I say all generators, I mean just that. All generators, whether they are IPPs, utility affiliates, utilities or others, will benefit from the lessons learned in our emerging competitive marketplace. As many in the U.K. market say, a winner is a company that is willing to learn and adapt, whether it is a utility or an independent.

Successful generators have learned to focus on those areas where they *can* add value. Generators must manage risks, push design technologies to the limit, and provide flexibility to our utility customers.

- For example, managing risks means, especially in the initial stages, limiting permitted risks. One way to minimize risks is to gain the support of local communities. Successful companies recognize that this means being able to go into a local community and earn the trust of community leaders, activists and residents. Please note that I am using the verb, "managing," very carefully in this context. In a competitive environment, controlling one's exposure to risk means knowing what risks to hold and what risks to transfer to those who are better able to manage them. This is one important way to keep costs to a minimum.

- Pushing new design technologies means being willing—as USGen was when setting up its East Syracuse Generating Plant here—to adopt new technologies like selective catalytic reduction (SCR) for reductions of nitrogen oxides emissions. This was the first use of SCR technology in a natural gas plant in New York. Initially, the cost was high but acceptable because SCR would give us the results we wanted. In an excellent example of how a competitive marketplace works, when USGen went out to get SCR units for its later natural gas plants the cost had dropped substantially.

- Third, and most important, generators add value by providing customers with new flexibility. This requires that they truly listen to their customers. The way generators respond is transforming the generation industry in subtle and not so subtle ways. For example, in its infancy, the IPP industry relied on the "PURPA put," or avoided-cost strategies to compel utility purchases—something that we, who do business in New York, are familiar with. You well remember when IPPs had little incentive—and showed little willingness—to address specific customer needs. Times have changed. We have stopped talking and begun listening to our customers. All of this listening means we have to tailor and adapt our power projects to suit those customers.

117

It is a given that a project will meet the customers' megawatt needs, but now we also have to consider the timing of the delivery of those megawatts, i.e., dispatch; the pricing of different kinds of megawatts; and the customers' risk appetite.

We at USGen have found from listening that customer needs vary widely. Some customers value dispatch flexibility. Some want local community concerns to be addressed, thus ensuring a project will be finished. Others find it desirable to retain interest-rate risk and manage it themselves. Still others value the opportunity to invest in the project if it is successful.

Flexibility—that is, the question of how the generator and the customer interact—may also be a relationship issue. Many utilities want to know that contract pricing and risk distribution can be restructured over time if situations change. In USGen's experience, contract restructurings have provided opportunities for both USGen and its customers to benefit. This adds value and strengthens customer relationships—a win-win-win proposition. We win, our utility customer wins and the ultimate customers, the ratepayers, win.

Generators' Business Approach—How Do We Do This?
Therefore, as generators we have and we will continue to add value. Let's look now at how this affects our approach to this business. The competitive generating marketplace is demanding that we do things differently than we have in the past. One obvious example is the shift from "PURPA put" to customer responsiveness. This has required people and organizations to change significantly. Another is the recognition that creativity and innovation are now necessities. Let me give you an example of how creativity and innovation come into play in our business.

In the uncompetitive market, the price of a new facility was reached by totaling up the cost of that facility's equipment and machinery. However, in the competitive marketplace, purchasers not suppliers—set the price. And over time, competition drives the price of our products down.

So focusing exclusively on cost—and how to reduce each element's cost—is not the key to surviving in a competitive market. The key to competitive survival is to produce a product that brings value to a customer. This value may be reflected in many ways, including price, project structure, timing, etc.

Chapter 9: The Electric Industry in Transition

One way to understand the principle that price should not be viewed as the simple sum of component costs is through an analogy to basketball. Let's say you're a basketball coach. Your team is playing a team with a 7-foot center who dunks like Shaquille O'Neal, two power forwards—each with better outside shots than your forwards, and a point guard who's quicker than yours. At first you think your team cannot win because it is outmatched at every position. So you don't want to match up with them one-on-one, but your team can win by playing smarter, playing more creatively and playing more aggressively.

- Playing smarter in our business means studying your competitors and learning where their weaknesses are and where one can better add value for the customer.

- Playing creatively may mean structuring a financing scheme to produce a more efficient solution to a customer's needs.

- Playing aggressively means always pushing for the next level of efficiency and productivity. One is never complacent, even after a win.

In the generation industry, we cannot get trapped in whatever other reality may seem to exist. We don't arrive at a price simply by adding up project costs. We play between the seams. We play smart. We play creatively. And we play aggressively.

Characteristics of the Current and Future Competition

As you can see, I envision benefits as we proceed to market efficiency in generation and transmission. The market will decide who the winners and losers are, but there is no question that new players will thrive in the competitive generation industry. The 1992 Energy Policy Act cleared the way for utilities and utility affiliates to enter the fray through the Exempt Wholesale Generator section. We now see power brokers and power marketers emerging—all joining the regulated utilities, IPPs, and qualifying facilities (QFs) already in the market.

The economic slowdown created by the 1990 recession depressed the market for new baseload capacity and created oversupply in some regions—most notably here in New York. The result today is a very competitive—that is, low-priced—short-term capacity market.

That's where we are today—many players in a depressed market. Looking to the future then, we are all asking the same question—whether this short-term capacity market is a temporary or a long-term reality. Frankly, I don't know. And it shouldn't matter—providing generators constantly do what has been proven to be successful, because what has proven to be successful will work in either regime. Look for new ways to add value; focus on customer needs; and be creative to avoid getting trapped by our old ways of doing business.

Generators who believe that future generating plants will be structured financially just as they are today, that is, by long-term contracts supported by single-asset facilities, are being trapped by our old ways of doing business. The commitment by an independent generator to go forward with the construction of a merchant plant would be a major breakout of this conceptual envelope. By a merchant plant, I mean a facility that does not have all of its electricity product sold under contract for a major portion of the project's life. Funding such a project would be difficult but it does represent one way to address the movement towards a short-term capacity market.

Let's take out the crystal ball and summarize where all this may be leading.

- ◆ We will have some merchant plants selling under both short-term and long-term contracts.

- ◆ Generators will, I believe, be selling mostly on a wholesale basis—and potentially to power brokers and marketers if those brokers and marketers can add value to the ultimate product.

- ◆ All generators will be free of regulation, that is, they will be free to compete without constraints. Indeed, to be free of regulation, generation has to be separated from transmission and distribution.

- ◆ Ownership of generation assets will be different. Pretty soon, the terms "hybrid" and "merchant plant" will be as common as "qualifying facility."

- ◆ New electric generators will be spun off from utilities, there will be some consolidation of IPP companies, and there will be some combined IPP and utility generation entities.

Chapter 9: The Electric Industry in Transition

The new generating company that finally emerges will not be an IPP or a generation division of a utility. It will be a new and unique entity born out of the successes of all generation players now in the game. I'm sure, also, that at the end of the process the broadening of competition will bring its usual results: lower end-use prices for customers and thinner margins for all players.

The costs and benefits of market efficiency will be a function of making all that we do relevant to the brave new world of competitive power generation and transmission. Generators have compiled an impressive record of success by adapting to change. Competition in the future will mean reapplying those lessons: being externally customer-focused and being smart, creative, and aggressive in selecting partners, structuring responsive projects, and directing our organizations to implement these creative and aggressive strategies.

Provided we make this effort, I believe we can compete successfully and our customers will be the beneficiaries of that success.

Part 4

The Electric Industry in Transition

INTRODUCTION

The Role of Regulators in Managing the Transition to Competitive Power Markets

James Brew
Assistant Counsel
New York State Public Service Commission

Developments in the electric industry in the last decade have set the stage for competitive power markets. The policies of rate regulators are expected to be crucial to the transition from cost of service regulation to market pricing of electricity. Regulators may believe that they can manage the pace or direction of the transformation of power markets, but this may not be entirely accurate. Actions that are being taken or proposed almost daily are altering the landscape in the industry.

State regulatory commissions throughout the country have initiated proceedings to consider these issues. It is not clear whether these proceedings will shape the course of the industry or confirm changes that have already occurred.

The authors in this chapter offer the views of industry observers who represent neither regulators nor regulated utilities. They include an electric industry investment banking advisor, a representative of large industrial customers, and a consumer advocate. Together, they provide thought-provoking perspectives on the roles regulators should assume.

Leonard Hyman, a former first vice president at the investment banking firm of Merrill Lynch, maintains that the role of regulators should—and will—diminish as the discipline of the marketplace asserts itself. As vital as electricity may be to modern society, it is a commodity, Hyman explains. Government oversight of power markets will be required to prevent anti-competitive practices and other abuses, just as those safeguards—generally provided by the Department of Justice, the state attorney general and the courts—are required for other industries and services.

Hyman argues that regulation of the generation sector will be increasingly limited as genuinely competitive markets develop. A residual oversight role may remain to address key concerns (e.g., fuel diversity), but it probably will be very different from the heavy role regulators now have in overseeing power production. Hyman notes the overlapping and competing demands of state and federal regulators over transmission and that there are several paths the industry could take. He observes that the transmission system could temper the transition to competition, particularly if transmission assets were deregulated or regulated on a fair-value basis. He also asserts that local distribution may be the last remaining monopoly but that options could be explored to make distribution more efficient.

Hyman concludes that electric supply is moving towards an open system and that regulators should assume the roles of referees or facilitators for the transition period rather than cling to past models of regulatory oversight and control.

Dr. John Anderson, Executive Director for the Electricity Consumers Resource Counsel (ELCON), argues that to a large extent competition already has arrived in the electric industry and is driven by technological developments and economics. He asserts that the industry will experience a chaotic period in the short term, but that expanded choice of suppliers for all customers ultimately should be allowed. He claims that the growing disparity between electric rates and the cost of alternative sources of supply have provided large industrial customers with a strong motivation to seek alternatives.

Anderson recognizes that the transition issues will be difficult and asserts that regulators will need to manage the competing tensions among industry participants (core vs. non-core customers, utilities vs. independent power producers (IPPs), inter-utility affiliate transactions, and disputes among utilities as competitors). He offers eight principles for a transition to a competitive electric industry that would allow for retail competition. He calls on regulators to seek an equitable means of splitting legitimate transition costs among ratepayers, shareholders and taxpayers, noting, however, that assets that were not economic under existing regulatory regimes should not be included in transition costs.

Anderson also argues that retail competition will eliminate utility-implemented integrated resource planning and demand-side management programs. Regulatory oversight in these areas, therefore, will need to be substantially revised to accommodate a transition to competition in the electric sector.

Introduction: The Role of Regulators in Managing

Dr. Mark Cooper, Director of Research for the Consumer Federation of America, maintains that regulators need to balance the competing interests of promoting competition in the electric sector with protecting core (also known as "captive") consumers. Key tasks for regulators, according to Cooper, will be to overcome utility resistance to eliminating bottlenecks to service such as transmission, and to prevent abuse of costs and strategic advantages resulting from an incumbent monopolist's position.

Cooper explains that regulators must recognize that competition does not come to all customers at the same time and that it may never come for some customers. He argues that regulators must ensure that captive customers are treated fairly and that public functions are preserved.

Cooper focuses on structural protections needed to prevent cross-subsidization of competitive services by captive customers and reallocation of risks. He also asserts that writedowns of some uneconomic utility assets are inevitable and that utility accounting treatment will need to be revised. He points out that local utilities should retain their obligation to provide service in high-cost areas, that common carriage principles may need to be applied to transmission services, and that reliability and quality of service for all ratepayers will need to be preserved. He also suggests that utility shareholders have been compensated through risk premiums and allowed rates of return through the years for investments made in plants that could be uneconomic in a competitive environment.

Cooper maintains that since investors have been compensated already for the prospect that costs might be stranded, if stranded cost recovery are to be considered, a retrospective review of risk premiums allowed in the past may be in order.

CHAPTER 10

The Role of the Regulator in an Increasingly Competitive Market: The Smaller the Better?

Leonard S. Hyman
Consultant and former First Vice President of Merrill Lynch, Pierce, Fenner and Smith, Inc.

Competition breeds efficiency; roots out the unfit; and, by providing choices, protects the consumer from exploitation. But in certain industries—styled as public utilities—competition would induce inefficiencies and lead to higher prices for the public. For these industries exhibit economies of scale. One supplier, encompassing the entire market, can produce its output at a cost far below those of many smaller suppliers competing against each other. That single supplier, with a monopoly over the market, uninhibited by competition, might take advantage of its position as sole supplier to overcharge its captive customers. In order to protect the public, the government sets up a regulatory agency whose purpose is to act as a surrogate for the competitive forces that normally would force the supplier to run efficiently and that would protect the consumer.

This process of regulation worked well for decades. Public utilities increased in efficiency as they grew larger, while the regulators assured that the consuming public derived benefits from those efficiencies through declining prices of service.

In the 1960s, technological changes began to erode the advantage of the natural monopolist. By the 1970s, new entrants started to nibble around the edges of regulated monopoly industries. In the 1980s, new laws encouraged competition as a substitute for regulation wherever possible. Deregulation took place in transportation. Wholesale utility industry restructuring in the United Kingdom and in the American telephone industry demonstrated that rigid regulation and integrated systems were not necessary to assure reliable service and reasonable prices.

With the fall of communism in this decade, capitalism seems to have triumphed, and with it the belief that market mechanisms of discipline are better than those devised by government agencies. In the case of the highly-regulated electric utility industry, most observers would agree that the regulatory system has not functioned well in the past two decades and that competition should be introduced as a substitute for regulation wherever possible. That process of substitution might lead to a situation in which certain sectors of the business remain regulated while others operate in a relatively free market. What role, then, does the regulator play?

- In which sectors should regulation remain? Presumably in those that retain the characteristics of natural monopolies.

- Should the regulator function as before in the sectors that remain regulated? Given the lessons learned both in the U.S. and abroad, one could conclude that new regulatory formulas might encourage greater efficiency and innovation than current arrangements, and should be tried.

- Should the regulator—as well—serve as the Food and Drug Administration or Fair Trade Commission of the electric supply industry, to assure that competitive forces work? Or should that function fall to the Department of Justice, the state attorney general and the courts, as it does in other industries?

As competition increases in the electric utility industry, the role of the regulator—eventually—will diminish. The disciplinary forces of the marketplace will then replace the curatorial and pedagogic rule of the regulator. Paradoxically, the regulator's role may enlarge temporarily in order to assure that the industry makes the transition to fair competition, although, in due course, the role of the regulator will shrink to that of guardianship over only those sectors of the market that remain natural monopolies.

Is Electricity Different?

That question, in turn, leads to a question that enters into the discussion almost in theological terms: is electricity different? If it is, supply of electricity may require a continued unique regulatory apparatus rather than reliance on the forces of the marketplace to achieve satisfactory performance. A priesthood of regulators, dedicated utility employees, appliance manufacturers, and consummate public relations personnel has formulated and encouraged a cult of electricity. Think of the icons: Benjamin Franklin, the kindly discoverer; Thomas Edison, the personification of American ingenuity; Charles Steinmetz, the second wizard of lightning; David Lilienthal, the proponent of electrification as a developmental force; Woody Guthrie, balladeer of hydropower; Ronald Reagan, salesman of eternal verities; Betty Furness, bringer of the home appliance into every living room; and of course, Reddy Kilowatt, the friendly spirit of electricity.

Electricity, we are told, is different. It is too important to society to leave in the unshackled hands of the free market. For without electricity, our society would cease to function. That is electricity seen as a religion, rather than as an energy source or a commodity. Believers demand continued regulation. They fear that some members of society could not acquire a sufficient supply of electricity at an equitable price. They fear that a free market might not supply the requisite electricity for the country's development. They fear that unregulated suppliers will choose the wrong technologies. They fear that unregulated suppliers will not support desirable societal and environmental goals. The discussions take on distinctly ideological slants.

Economic Motives

Perhaps, though, one should attempt to delve beneath the public rhetoric to seek economic motives. When competition enters a slow-growing market, the new entrants may take business away from the incumbents rather than simply serve the new demand. In other words, the competitors not only take market share but they take market as well. If, moreover, the old marketplace featured cross-subsidiaries, those consumers paying the subsidies will seek to take their trade to the competitors whose prices do not contain subsidy elements. Society as a whole may benefit from competition, but all participants do not. Deregulation and competition could affect adversely many groups that benefit from today's regulated structure.

◆ Numerous electric utilities would lose revenues if lower cost competitors could reach their markets. Creditors and shareholders of these utilities, who placed their faith in a dissolving regulatory compact, could incur losses of income and principal. Those utilities may require the protection of regulation for an extended period of time, while they reduce their costs to competitive levels and, if necessary, write down uneconomic assets in an orderly fashion.

◆ Independent power producers, if they had to deal directly with consumers, would lose comfortable competitive and financial advantages. Independent power producers could no longer rely on a long-term power purchase contract signed by the utility as the sole customer to provide the necessary backing for financing. They would have to deal with numerous potential customers, many of whom might not be willing to sign such extended contracts. That may account for the reluctance of independent power firms to espouse real competition at the ultimate customer level.

◆ Consumer groups (whether representing large industry or small users) know that rebalancing of prices will take place when the cross subsidies unravel or when those with the greatest bargaining power flex their muscles. Rebalancing may mean that large users end up paying less and small users end up paying more. If so—whatever the justice of the change in prices—representatives of the potential winners will advocate greater competition and representatives of potential losers will resist.

◆ Environmental and social activists have fashioned a number of programs whose costs are incorporated in the price of electricity. The continuance of those programs may depend on preventing the consumer from choosing whether to fund those efforts or not. The competitive entrant may not pay to support the programs, and can, therefore, charge less. Consumers—especially business firms competing to sell their own products in a worldwide marketplace—tend to opt for low cost suppliers, if they can. Those consumers want social and environmental programs funded though taxation that falls on all their competitors. Freeing up the electricity market could endanger those programs as they presently exist, although pricing signals could substitute for command-and-control in a properly designed competitive system.

◆ Governments use utilities as tax collection agencies. The tax burden placed on the utility may not be as easily placed on multiple suppliers and self generators. Thus, opening up the utility market may affect a significant source of state and local governments' revenue.

◆ Finally, an army of regulators and the assorted camp followers, the lawyers and consultants and expert witnesses, would have less to do, in a less regulated environment, thus fewer jobs.

Although theorists and potential winners see economic benefits from greater competition in—and less government regulation of—the electricity supply industry, those who foresee economic losses to themselves or to the programs that they favor could raise a formidable opposition to the process.

Chapter 10: The Role of the Regulator

Electricity Supply as a Commodity Business

One can demonstrate easily that food, another necessity, remains on the market in our state despite the fact that we lack a New York State Food Service Commission. Consumers can choose from expensive or inexpensive products. One can assert smugly that poor people can acquire food by using food stamps, so we have combined the benefits of the abundance of a free market with the virtue of a social safety net to help those in need. Unfortunately, a significant number of people go hungry. The effort is not a universal success. One solution to that problem would be to control every aspect of food production, pricing and distribution: perhaps charge premium prices for muesli, for instance, in order to subsidize corn flakes, or, even better, institute income requirements for entrance to stores to make sure that bankers driving the inevitable BMW do not shop in the low priced stores reserved for domestic workers who take the bus. Government could establish special training programs to help people to eat the right foods for their better physical and financial health. The alternative is to accept the workings of the marketplace and do a better job of providing for those in need. Occam's Razor or Rube Goldberg? That is the choice.

Electricity supply, in the British sense of the term, means the acquisition and transfer to consumers of a quantity of electric power. The electricity itself is a commodity. Competing generators produce it for sale, and the most efficient generators make the greatest profits. The electricity is the same everywhere, but quality of service differs. Consumers should be willing to pay different prices for different levels of quality if so offered. Even so, regulators may need to set and police certain minimum standards so that the entire system can work as a coordinated whole. To the extent that some consumers do not have choice of supplier, the regulators must assure that the agents who procure the supply do so at a reasonable price and that they avoid conflicts of interest that disadvantage the captive customers. Their role, though, could diminish if entrepreneurs form virtual electric companies akin to long distance telephone resellers, in which case competitive forces might prevent exploitation of customer groups.

Karl Marx wrote: "The battle of competition is fought by cheapening of commodities." Producing the electric energy accounts for more than one-third of the price of electricity. Recognizing electricity as a commodity, and then encouraging competitors to drive down the price, may produce the biggest payoff for consumers.

Regulation of Generation/Supply
When generation demonstrated economies of scale, regulators barred the entry of competing generators in order to assure that the designated local monopolist could attain the maximum economies of scale. That tactic assured consumers of the lowest cost of power. Economies of scale ceased in the 1960s. The development of the gas turbine, combined with low natural gas prices, created an environment in which small, easily-installed generating units could compete against the leviathans that typified the utility industry. In effect, not only did natural monopoly—and the rationale for regulation of generation—expire, but so did the barriers to entry that would have protected the market position of the incumbents.

If the transmission and distribution systems are open, so that generators can reach their customers, regulators could play an increasingly limited role in the generation sector. Eventually, regulation (except in the antitrust sense) of a fully competitive generating and supply industry could cease. It would remove regulation from half the assets of the electricity supply industry. That would raise questions, too. Most new generating plants burns natural gas. What if gas prices shoot up? If national policy requires the installation of more nuclear power plants, how could one force that decision in an unregulated market? Who would force the market to take into account environmental needs? Could one trust generators to make decisions that would correct environmental abuses as well as line their pockets? What if new technologies produced such economies of scale that consumers would benefit from a return to regulated natural monopoly? Those are not trivial questions, but it should be noted that there is little evidence that governments or regulators have done a better job of predicting the future or protecting the environment than market-driven firms.

Perhaps the answer is that regulators should retain residual authority, even in the competitive sector, but not act unless events require a resumption of regulation. Even then, the regulators need not play the same role or utilize the same methods as they do today.

Regulation of Brokering and Pooling
In a competitive system, certain entities will, as businesses, acquire and dispose of electricity. Some will operate the power pools and regional control centers that will deal with the physical issues involved in running a reliable, integrated and efficient electric supply system. Some will transport the power over long distances. Unless the brokers engage in

pervasively dishonest activities or corner the market in this vital commodity, brokerage should require the same type of regulation as the sale of any commodity or security. Essentially, that, is regulation that assures that the participants follow the rules of the game.

Regulation of Transmission

According to prevailing opinion, the transmission bottleneck thwarts true competition, and transmission is a natural monopoly that requires regulation. Not only is it regulated, but it is enmeshed, Laocoon-like, in a regulatory gridlock created by multiple regulatory agencies all asserting their primacy. The states want to protect native load customers. The FERC, which wants open transmission, encourages endless use of a finite resource. Does the system have a strong resemblance to and promote the same consequences as rent control? Some people have exceptional apartments at low costs regardless of their income levels; landlords hesitate to upgrade rent-controlled assets unless the upgrade produces a bureaucratically-approved price increase; and tenants literally inherit the rent-controlled apartment despite the fact that the landlords—supposedly—own them.

According to the prevailing wisdom, the transmission system is a finite resource. It is almost impossible to expand the existing transmission system. Therefore, that resource requires strict regulation in order to prevent the monopolistic owners from taking advantage of the consumers. Consider as an alternative the possibility that utilities are reluctant to expand the transmission system because it is so regulated, because transmission services are priced in a way that often leaves owners uncompensated for the use of their systems, and because the transmission system seems to the FERC more of an instrument of public policy than a business venture.

Regulators can deal with transmission in several ways.

♦ They can continue to regulate it in the same way as they do now, hoping that the Electric Power Research Institute's efforts to expand the workability of the existing network bear fruit, and that owners of the system will find a reason to install the new equipment that expands and improves the system.

♦ They can devise a new system of regulation that provides the owners an incentive to expand the system's capacity, whether by adding new lines or new equipment, or by bringing about more rational economic use of the system through pricing policies.

The Electric Industry in Transition

- They can remove transmission from shared ownership with other aspects of the electric supply industry, to end all discussion of who paid for what and who is entitled to first call on the system, thereby obviating the issue of cross subsidy, but possibly losing control to one regulatory agency, the FERC.

- They can transform the transmission system into a governmentally-owned entity akin to the highways or air traffic system and possibly lose all control.

- They can decide that the transmission systems provide alternate routes, as do railroads and airlines, sell them off to the highest bidders, and step back from regulation.

The transmission system could play a tempering role in the transition to competition, although the likelihood of that happening seems slight given the plethora of stakeholders seeking use of those limited assets. If transmission were freed of regulation, or if it were regulated either on a fair-value basis, or on some other basis other that which exists today, the transmission entities might become sufficiently profitable to entice buyers to pay a price for them high enough to produce a profit for the sellers that could offset some of the losses derived from writing down uneconomic assets and from buying out uneconomic power purchase contracts. Access fees to the transmission system could, too, cover the costs of some of the social and environmental programs that might otherwise disappear in a competitive environment.

Regulation of Distribution
The local distribution system may constitute the last remaining natural monopoly. Until an interloper demonstrates that it can lay duplicate electric lines at costs that make the effort worthwhile, most suppliers of electricity will prefer to utilize the one existing system. (The system, though, does not have complete control of the market. Distributed generation, for example, is an alternative to more distribution plant. Presumably, neighborhoods could attempt bypasses to tap the transmission system, too, if the distribution system charges exorbitantly.)

The fact that the distribution apparatus likely will remain under the guidance of regulation because of its monopoly status leads to several problems.

Chapter 10: The Role of the Regulator

- Regulators may be tempted to load uneconomic social and environmental programs onto the distribution customer, which could tempt some large consumers to find a way to bypass the distribution system.

- They may try to control and regulate service offerings over the distribution system that could be supplied profitably on a competitive basis. Doing so would hinder the development of services that should emerge as the electricity market fragments.

- They may turn the existing distribution utility into a supplier of wires, keeping it out of all competitive service offerings made over those wires, for fear that the utility would dominate those offerings. Doing so may keep out the one firm that has an interest in serving a broad market.

- The local distribution utility, forced to act as supplier of last resort if the competitive market does not serve some segments of the population, may have to take on supply and contractual obligations for which it will not be compensated.

Even if distribution does remain regulated, the regulatory system need not work on the assumption that nothing can make distribution more efficient. New regulatory formulas that encourage greater efficiency and innovation can be applied to the distribution sector.

What Is The Role of the Regulator?

Once upon a time, the electric supply industry was a closed system, or at least as closed as could exist in the economy. The regulator could influence outcomes, possibly even determine outcomes. The regulator could set technological and social agendas. (Some states not only specified the technologies that the utilities had to choose, but also ordered the prices that the utilities had to pay for those technologies.) The utilities executed the agendas. The consumers derived benefits from or suffered the consequences of the agendas. Leakage from the regulated system was minimal. Customers might use less electricity, turn to natural gas (another regulated industry) or to other fuels, or even move away; but if they wanted to use electricity where they lived, they paid the bills. When electricity was cheap and getting cheaper, one gained little from jumping ship.

The Electric Industry in Transition

The electric supply sector is moving toward an open system. The regulator has less control of outcomes. Disaffected consumers develop means of escape through retail wheeling, wholesale wheeling, self generation or a move to a different state or country. Technology and regulation create leakage. (The customer can install efficient gas turbines, rather than buy from a utility. The FERC has opened up access to transition lines, so customers can tap sources that are reachable through the lines.) Eventually, the water pours out, creating a vision of the electric supply industry that resembles the leaky boat in Laurel's and Hardy's "Towed in a Hole." The regulator, at that stage, has to choose a role, possibly that of the Little Dutch Boy (the heroic mode), King Canute (the legal mode), or Stan and Ollie. Neither Lord Nelson nor Admiral Farragut are among the options.

What should the regulator do? Leaving aside the naval metaphors, the most productive role for the regulator would be that of referee and facilitator during the transition period, designing and implementing procedures that bring about the most orderly and least painful movement toward an open system, taking care to encourage desirable social and environmental ends through the use of market-based pricing mechanisms, and supervising the transition so that a fair, open, and competitive marketplace actually develops. After that, the regulator will supervise generation lightly, and transmission and distribution more heavily. Where regulation continues, the regulator needs to develop procedures that encourage innovation and efficiency. Clinging to the prerogatives of the past is not a productive option.

References

1. California Public Utilities Commission, *Ordering Instituting Rulemaking and Order Instituting Investigation*, R.94-04-031, I.94-04-32, Filed April 20, 1994.

2. Comisión Nacional de Energía, *El Sector Energía en Chile* (Santiago: Comisión Nacional de Energía, December 1993).

3. Dasovich, Jeffrey, William Meyer and Virginia A. Coe, *California's Electric Services Industry: Perspectives on the Past, Strategies for the Future, A Report to the California Public Utilities Commission by the Division of Strategic Planning*, February 3, 1993.

4. Falcone, Charles A., "Retail Wheeling...The Sword of Damocles," presented at the Energy Daily Conference on Retail Wheeling '94, March 8–9, 1994, Arlington, VA.

5. Henney, Alex, *A Study of the Privatization of the Electricity Supply Industry in England & Wales*, (London: EEE Ltd, 1994).

6. Hyman, Leonard S., *America's Electric Utilities: Past, Present and Future* (Arlington, VA: Public Utilities Reports, 1994).

7. Niagara Mohawk Power Corporation, *The Impacts of Emerging Competition in the Electric Utility Industry*, April 7, 1994.

8. Pacific Gas and Electric Company, "Competitive Positioning," April 1994.

9. U.S. Congress, Office of Technology Assessment, *Electric Power Wheeling and Dealing: Technological Considerations for Increasing Competition*, OTA-E-409 (Washington, D.C.: U.S. Government Printing Office, May 1989).

10. TransAlta Utilities Corporation, *Proposed Rates to Respond to Evolving Markets for Electric Services in TransAlta's Service Area*, December 1992.

11. Rochester Telephone Corporation, *Petition of Rochester Telephone Corporation for Approval of Proposed Restructuring Plan*, Case No. 93-C- , February 3, 1993.

CHAPTER 11

The Role of the Regulator in a Restructured Electric Industry

Dr. John Anderson
Executive Director
Electricity Consumers Resource Council

Competition not only is coming to the electric industry—to a large extent, it is here. Competition is driven both by technological development and economics. The result, at least in the short-run, will be a mess. There is a better way. Let all customers competitively source their purchases of electricity.

The electric industry has been subjected to a healthy dose of competition over the past decade. It will become even more competitive in the near term with or without additional legislative or regulatory initiatives. The critical question is: in what form should competition be structured to assure fairness and equity for all stakeholders?

Competition in the electric industry is coming for several reasons:

♦ Both the level and the magnitude of the disparity in electricity rates have grown significantly over the past two decades. Specifically, the highest published industrial rate in the early 1970s was lower than the lowest today. Perhaps of even more importance, the spread from the lowest to the highest industrial electricity rate in the early 1970s was only about 3¢; today that spread is over 10¢. Similar increases in rate differentials apply to other customer classes.

♦ Most utilities today charge customers far more than the cost of new supplies. Specifically, electricity today can be generated for approximately 3¢ per Kwh, yet customers are usually charged multiples of that amount in their retail rates.

♦ Under these conditions customers have strong motivations to seek lower cost alternatives. The potential benefit—and thus the motivation to shop—is great. For example, a "typical" industrial often pays a penalty of several million dollars per year at a single plant site compared to a competitor located only a few miles away.

- Increasingly, industries have options freeing them from captivity to the local utility. Examples include (to mention only a few):

  - Self- or cogeneration

  - Inside- and outside-the-fence exempt wholesale generators (EWGs)

  - Dispatching or shifting production

  - Distributed generation

  - Municipalization (and reverse municipalization)

  - Retail-to-wholesale load shifts

  - Wheeling to political subdivisions of states

  - Jumping franchises

  - Utility-brokered power and "real-time" pricing

- The EPAct increased the viability of several of these options, giving an extra boost to retail competition. For example, the EPAct:

  - Made self-generation more viable by allowing the owner of a generator to sell excess power as an "exempt wholesale generator" (EWG). Prior to the EPAct, any sale of power made the owner of the generator a public utility subject to regulation. Now, EWGs can be made of a technologically efficient size, rather than designed merely to match the industrial load.

  - Assured municipal electric systems access to transmission, thus making municipalization more advantageous.

  - Allowed the Federal Energy Regulatory Commission (FERC) to order wheeling to ultimate customers who are served by "political subdivisions of the state," a very broad category in which customers are allowed widespread latitude to obtain the equivalent of retail wheeling through non-sham transactions.

Chapter 11: The Role of the Regulator

The electric industry will experience significant changes as these options are implemented. There will be:

◆ Many new entities supplying power (including power marketers, brokers, EWGs, qualifying facilities (QFs), municipals, etc.)

◆ Many new entities shopping for power (including power marketers, brokers, new municipals, political subdivisions of states, ultimate customers, etc.)

◆ Sizeable so-called "stranded costs"

◆ A plethora of special rates—each designed to stop the bypass

And all of these will occur without retail wheeling!

An unstructured movement (or, worse, stampede) towards retail competition will not be a pretty sight. What do I mean when I say conditions in the short-term will be a mess? I list several.

◆ Many independent power producers (IPPs/QFs/EWGs) will compete with utilities to supply energy and power.

◆ Municipals, co-ops, and political subdivisions of states will actively shop for power.

◆ There will be a growing use of power brokers and marketers. Indeed, there now are more than 40 marketers now approved by the FERC. This number will continue to increase.

◆ Customers of all persuasions will attempt to exercise all available options.

◆ Utilities will discount heavily, making special deals with many customers, in an attempt to stop the bypass.

◆ Wall Street will downgrade many utilities. Security analysts will carefully scrutinize high-cost utilities that present a credit risk to stockholders, as they do other industries.

◆ High-cost utilities will experience sizeable "stranded costs" as customers find ways to bypass uneconomic assets.

The Electric Industry in Transition

◆ Many utilities will downsize (or "rightsize" as those in other industries term the reorganization). The downsizing will be much greater that any experienced by electric utilities to date.

◆ There will be a growing pressure to eliminate social programs implemented through electric utilities.

These conditions will create significant tensions between various stakeholders in the electric industry. Examples of such tensions include:

◆ Core vs. non-core customers. Some customers will receive special rates or discounts. Remaining customers will be very concerned that they will have to pick up so-called "stranded costs." Significant discussion will center around return rights.

◆ Vertically-integrated utilities vs. IPPs and consumers. Both IPPs and consumers will be very concerned about self-dealing and preferential treatment regarding subsidiaries such as utility-affiliated generating divisions and fuel suppliers. These concerns will relate to both domestic and foreign subsidiaries.

◆ Utilities vs. utilities. Utilities that in the past would cooperate with each other will now consider their neighbors as potential competitors. This may restrict voluntary cooperative efforts. As an aside, the increased level of competition may make the operation of voluntary wholesale power pools quite inefficient when the participants include vertically-integrated utilities.

◆ State vs. federal regulators. Already, there are jurisdictional disputes between state and federal regulators. However, these disputes will greatly increase as the electric industry becomes even more competitive. Questions such as the following will have to be answered.

– What is retail service anyway?

– Is there such a thing as intrastate transmission?

– Who has authority over unbundled retail transmission?

– Who can or should approve the creation of new wholesale entities?

Chapter 11: The Role of the Regulator

- Is the wheeling to a political subdivision of a state retail or wholesale?

- Can non-transmission costs be included in transmission rates?

- Can exit fees be assigned to departing customers?

We truly will be living in interesting times. However, there is a better way. All customers should be able to source their power needs competitively.

Competitive sourcing—or, more narrowly, retail wheeling—can produce many significant benefits for all customers.

◆ The planning process will be greatly improved. True least-cost planning will be assured. Planners will become risk-takers, thus making them approach the planning in a much different way.

◆ A benchmark is provided for those who choose not to shop. Regulators can compare the price charged to non-shopping customers to that obtained by shopping customers.

- Cost-effective energy efficiency measures will be encouraged. Retail competition will lower electricity costs, and thus increase economic activity. Since environmental protection increases with economic affluence, retail competition benefits the environment. Furthermore, customers are more cost conscious when they shop. Retail competition will heighten the knowledge, and thus the implementation, of truly cost-effective energy efficiency measures. Finally, retail competition will give customers a real way to vote for "green" resources. Energy efficiency advocates have said that customers want and will pay for green resources. Retail competition provides a true test of what customers really want.

- Some argue that so-called "stranded costs" are so large that retail competition cannot be implemented. This assertion is incorrect for at least two reasons. First, stranded costs may be significant, but they are nowhere near the $200–300 (or more) billion dollars often suggested. Besides, whatever the level of actual stranded costs, they are a real indicator of the value of retail competition. Costs are "stranded" because they are uneconomic. Retail competition will make it impossible for uneconomic costs to be foisted on consumers. The elimination of stranded costs thus constitutes, dollar for dollar, an economic stimulus to the economy.

Several states have taken specific actions to investigate retail competition. California and Michigan provide very visible examples, but they are only two of more than a dozen. The Electricity Consumers Resource Council (ELCON)[1] recognizes that the transition to a competitive market will not be easy. We thus compliment those states that have recognized the values of retail competition and are taking the bold and courageous steps necessary for implementation.

ELCON offers eight principles as a way to minimize the objections and hitches.[2]

Principle N°. 1. Market forces can do a better job than any government or regulatory agency in determining prices for a commodity such as electricity.

Principle N°. 2. Laws and regulations that restrict the development of competitive electricity markets should be rescinded or amended. The need for burdensome regulation will be reduced when competitive electricity markets are allowed to flourish.

---

[1] ELCON was organized to promote the development and adoption of coordinated and rational federal and state policies that assure an adequate, reliable, and efficient electricity supply at competitive prices. ELCON member companies own and operate manufacturing and other facilities throughout the U.S. and in many foreign countries. ELCON member companies produce a wide range of products—including aluminum, steel, chemicals, petroleum, industrial gases, glass, motor vehicles, electronics, textiles, paper products, and food. Combined, the 22 members of ELCON consume over four percent of the total electricity in the United States. Many ELCON members cogenerate or generate some of their electricity requirements. Dr. Anderson is ELCON's executive director.

[2] These principles are contained in the publication titled *Retail Competition in the U.S. Electricity Industry — Eight Principles for Achieving Competitive, Efficient and Equitable Retail Electricity Markets*. This publication is available from ELCON upon request.

Chapter 11: The Role of the Regulator

Principle N°. 3. The benefits from competition will never fully materialize unless and until there is competition in both wholesale and retail electricity markets; but not all retail electric services are natural monopolies and therefore they should not be regulated as such.

Principle N°. 4. The owners and operators of transmission and distribution facilities and the providers of coordination and system control services should be required to provide access to those facilities and services to any buyer or seller on a nondiscriminatory, common-carrier basis.

Principle N°. 5. Rates for the use of transmission and distribution facilities should reflect the actual cost of providing the service. If the facility is a natural monopoly, those rates should be based on actual costs and the services provided on a nondiscriminatory and comparable basis to all users.

Principle N°. 6. Resource planning is not a natural monopoly. The types and market shares of generation and end-user technology that will be supplied in wholesale and retail markets should be decided in the marketplace.

Principle N°. 7. Legitimate and verifiable transition costs that develop as a result of competition should be recovered by an equitable split among ratepayers, shareholders, and taxpayers. The costs of assets that are uneconomic in the existing regulatory regime are not transition costs.

Principle N°. 8. The potential for transition costs should not be used as an excuse to prevent or delay the onset of a competitive electricity market.

I add a few thoughts on competition and the environment. It is often asserted that retail competition will have tremendously negative impacts on the environment. Specifically, it is claimed that retail competition will eliminate utility-implemented IRP and DSM programs, which in turn will eliminate the environmental benefits that are now being realized from these programs.

At the outset, ELCON emphasizes that large industrial electricity customers strongly support energy efficiency and are very concerned about the goals of social programs. They have to. Fierce domestic and international competition requires them to implement energy-efficient improvements wherever they are cost-effective.

However, we have serious problems with how energy efficiency and conservation programs (generally called demand-side management, or DSM, programs) are being implemented through electric utilities. We identify six specific points in our written testimony.

First, the implementation of DSM programs raises electricity rates. Electricity rates already are too high. The unnecessary increase punishes economic activity. The implementation of retail competition will lower electricity rates, increase the productivity of the electric industry, and stimulate greater economic growth.

Second, high electricity rates unduly burden low-income citizens by: (1) forcing them to pay high electric bills that they can ill afford to pay, (2) increasing the prices they pay for electricity-intensive products and services, and (3) limiting job opportunities.

A recent article by Ronald Sutherland of Argonne National Laboratory provides convincing support for these points.[3] Sutherland's research was conducted for the Office of Policy, Planning and Program Evaluation of the U.S. Department of Energy. Sutherland analyzed the statistical association between household income and participation in electric utility energy conservation programs and the association between participation and the electricity consumption. He found that:

- No statistical associations between reductions in electricity use and participation in residential DSM programs.

- The highest income class has the largest participation in utility programs.

- The lowest income class has the lowest participation in utility programs.

- Utility DSM programs do save energy—but through the behavior of non-participants, not of the participants!

- In essence, low-income nonparticipants are subsidizing high-income participants. Nonparticipants are forced by higher rates to reduce their electricity consumption.

---

[3] Sutherland, Ronald J., *Equity Implications of Utility Energy Conservation Programs*, Argonne National Laboratory, Washington, D.C., February 1, 1994.

Chapter 11: The Role of the Regulator

- ◆ Participation was found to be positively associated with new residences, new water and space heating equipment, and the presence of insulation and other conservation measures. Because these dwellings are already energy efficient, the results suggest that utility programs may not save as much energy as expected, *i.e.*, the participants are free riders.

- ◆ The author concludes that high-income ratepayers participate not to save energy—they already are using energy efficiently—"but to make themselves better off" at the expense of other, low-income, ratepayers.

It is quite clear that the rate reductions that can be achieved in a restructured industry will provide low-income ratepayers with greater relief than that which is currently provided by special assistance programs offered through electric utilities.

Third, the affluence generated by economic growth increases the total resources a society is willing and able to devote to environmental protection and other social objectives. High economic growth will enhance our society's willingness to pay for environmental and other social amenities.

Fourth, central planning simply is neither as efficient nor as productive as market forces in cost-effectively achieving the goals of environmental protection and other social objectives. This is being demonstrated repeatedly in other sectors of the economy where market mechanisms are proposed as more workable alternatives to traditional forms of government intervention.

Fifth, we emphasize that attempts to achieve public policy objectives by using regulated electric utilities as "principal agents" for delivering energy efficiency or social services, by definition significantly reduce the purchasing power of all ratepayers. Indeed, up to 90¢ out of each dollar may be lost in the process in some DSM programs.

Sixth, several things can be done to increase the cost-effectiveness of environmental protection and other social objectives greatly relative to the results typically realized by programs administered under the traditional regulatory regime. These include:

The Electric Industry in Transition

- Considering all resources—not just some

- Accurately measuring all results

- Allowing neither lost revenue recovery nor incentives beyond actual costs

- Implementing laws—not trying to legislate new mandates

- Pricing all services according to actual cost of service

- Not attempting to internalize any externalities not mandated by law

CHAPTER 12

Protecting the Public Interest in the Transition to Competition in New York Industries

Dr. Mark N. Cooper
Director of Research
Consumer Federation of America

The following discussion is framed in general terms applicable to all network industries undergoing a transition to competition. However, after beginning with general concepts, the discussion takes the telecommunications industry as an example because it is much farther along the path to competition than the electric utility industry. The problems are briefly identified and then solutions to the most pressing problems are offered.

Regulators face two primary sets of difficulties in attempting to ensure that the introduction of competition into a network industry is in the public interest—promoting competition and protecting (residual or transitional) captive consumers. Promoting competition requires overcoming the resistance of the incumbent monopolist by eliminating bottlenecks and preventing the abuse of cost and strategic advantages that result from incumbency. Protecting residual captive ratepayers requires that commissions recognize that competition does not come for all customers at the same time and that it may never come for some. The commission must ensure that the remaining captives are treated fairly and that public functions are preserved.

In the first instance, the introduction of competition involves overcoming bottlenecks. These are services, terms and conditions that competitors require on an equal basis in order to be able to compete effectively. These conditions include both technical and business functions, as described in Table 12-1. Bottlenecks are particularly important in network industries.

Competitors must be afforded the same availability of these services at the same price that is imputed (or actually charged) to the incumbent monopolist on an unbundled basis. Unbundling is crucial to allowing

The Electric Industry in Transition

Table 12-1. Bottlenecks

A. Interconnection/interoperability
   1. Physical interconnection
   2. Technical standards
   3. Planning and timing of functionalities

B. Transport
   1. Right of way
   2. Pole and conduit space

C. System support
   1. Installation
   2. Maintenance
   3. Testing
   4. Restoration

D. Business Functions
   1. Marketing
   2. Billing
   3. Accounting
   4. Data processing
   5. Design and build support
   6. Research and development

competitors to find efficiencies in value adding. If the commission does not ensure equal access to bottleneck facilities or functions, then competitors will be stymied. Equal access requires technological and economic equality.

If the incumbent monopoly companies are to sell these functions or services to competitors, a fundamental problem arises as to how the revenues will be used. Each and every one of these functions is paid for today by captive monopoly subscribers. To the extent that revenues are raised by the sale of these services, then they must be credited to the benefit of basic service subscribers. Thus, revenues gained in creating the conditions for local competition should go to lower basic service rates. Ratepayers, who have granted the franchise to the incumbent monopolists and conferred the benefits of bottlenecks and their resulting market power on them, have every right to extract the maximum benefit from the sale of these elements.

Table 12-2. Cost and Strategic Advantages

A. Finance
1. Transfer of assets
2. Leveraging
3. Good will

B. Legal
1. Resale
2. Taxation
3. Franchise treatment
4. Certification

C. Scope of Entry
1. Technological
2. Product
3. Geographic

D. Economies
1. Size/capital
2. Scale
3. Scope

E. Strategic Behavior
1. First mover
2. Tying and bundling
3. Contracting

A second area of concern when promoting competition involves the cost and strategic advantages that the incumbents enjoy. The advantages of incumbency, described in Table 12-2, include factors such as access to customer lists, economies of scale and scope, and the ability to leverage residual market power to name just a few. These advantages, gained from having been a monopolist, can inhibit competition long after equal access to the network is granted.

Pricing of services to protect the public interest in the transitional and mixed competitive/monopoly environment requires that competitive services are neither cross subsidized nor priced in a fashion that unfairly

Table 12-3. Pricing for the Transitional or Mixed Enterprise

A. Cross Subsidy
  1. Incremental cost
  2. Stand-alone cost

B. Allocation of Risk
  1. Cost of capital
  2. Capital structure
  3. Leveraging

C. Allocation of Economies of Scale, Scope and Incumbency
  1. Joint and common costs
  2. Cost rules

D. Stranded Investment
  1. Writedown of assets
  2. Accounting treatment

disadvantages competitors (see Table 12-3). Ratepayers who remain the captives of incumbent monopolists (either transitionally or in the long term) may not be abused and must receive the full benefits of the costs that they are bearing.

Regulatory mechanisms to insulate basic service ratepayers from the risk of competitive enterprises must be established. Economic advantages gained by the incumbent should not go to the benefit of unregulated subsidiaries, but should lower the cost of monopoly services.

Certain functions and services in the public network possess the qualities of public goods (see Table 12-4). Competitors will not supply these services on their own and incumbent monopolists are going to be reluctant to do so if their local franchise is eroded.

The incumbent monopolists will continually complain that having to bear the cost of these public goods is hampering their ability to compete. They will try to shift the costs into isolated categories paid by ratepayers or recover them from competitors through allocation to access or other charges. Needless to say, ratepayers will resist the former attempt, while competitors will resist the latter. The commission cannot allow the quality of these functions to deteriorate, but they are not likely to be provided efficiently or at adequate levels through competitive solutions.

Table 12-4. Public Functions

A. Obligations
1. High cost areas/the obligation to serve
2. Common carriage
3. Reliability/quality

B. Services
1. Emergency service
2. Mutual compensation arrangements

The discussion of solutions that follows is grouped into four sections: regulation (including the analysis of it) commensurate with market power; regulatory structures (including structural protections and price rules); getting prices right despite cross subsidy and economic coercion; protection of the customers, which includes discussions on allocating the burden of joint and common costs, the reasons for minimizing the contribution from captive ratepayers, and rate rebalancing plus stranded investment; and public functions.

Analyzing market power. It has become popular to argue for "regulatory parity" between entities offering competing services. If the market power of the entities differs, however, regulatory parity is a prescription of disaster. Regulation should be commensurate with the market power of the entity regulated. The incumbent may need to be regulated, while the entrant does not. If the incumbent is deregulated prematurely, neither regulation nor market forces will discipline anti-competitive and anti-consumer behaviors.

Regulators should take a cautious approach to the prospects of competition diminishing market power. While competition is slowly increasing, competitive forces remain quite weak in many key segments of the industry.

The simple fact is that competition may come for more of the basic service segment; but, then again, it may not. It may come only in a decade, which will leave the company an immense window for mischief, should it be prematurely deregulated. With premature deregulation, the incumbent monopolist can continue to exploit its monopoly base for profit while it makes strategic and pricing moves to prevent competition from effectively entering the field. Neither consumers nor competitors are well-served by premature deregulation of pricing or profits.

The ability of incumbents to leverage their monopoly to prevent competitors from offering services and to monopolize services that depend on the local network service has been demonstrated time and again. Incumbents control the bottlenecks and public functions described above.

Manipulation of the price or terms and conditions of access to services give incumbents tremendous market power. There are additional cost and strategic advantages that incumbents enjoy.

It is absolutely crucial for analysts and regulators to recognize that just because competition is legal does not mean it exists. Even when a few competitors have entered the field, it does not mean competition is effective. Competition must exist before deregulation or pricing flexibility is granted—it does not exist today.

Tests of effective competition should include:

◆ Consideration of the number and size of actively participating alternative providers

◆ The extent to which directly comparable services are available from alternative providers in relevant markets

◆ The ability of alternative providers to offer equivalent services at competitive prices

◆ The market share held by the telephone company

◆ Whether the incumbent monopolist is earning excess profits on the service or product

◆ Whether the Commission has rules for reclassifying a service if it proves not to be competitive over time

Ultimately, effective competition means multiple suppliers for significant numbers of subscribers with significant numbers of subscribers having taken alternative service.

Regulating to reflect market power. Wherever monopoly power exists, full regulation must be preserved. As long as incumbents remain dominant players in the market and can exercise market power, regulatory

Chapter 12: Protecting the Public Interest in the Transition

safeguards must remain in place to protect consumers and competition. Relaxed regulation of competitive services is possible, but the predicate for relaxed regulation should be real competition.

There are typically two broad categories of solutions to the problem of integrated companies providing both competitive and monopoly services with joint and common costs—structural separations and accounting protections. Because the monopoly power of the network industries has been so pervasive and the services they deliver have such low elasticities of demand (i.e., are such vital necessities), both categories of protections are necessary during the transition to competition.

The discussion of regulatory structures may be divided into sections on structural design and price rules.

Where control over the switch or the network is crucial to the service, companies should give up the design and access monopoly. Design decisions should be subject to regulatory oversight, with input and review from competitors.

The problem of the cross-subsidization of non-telecommunications businesses must be addressed in traditional terms of separation of activities through subsidiaries. Complete separation between regulated and unregulated entities must be maintained. Assets, debts, property, employees, management, boards of directors and all operations should be kept separate.

Rules to restrict and regulate affiliate transactions to prevent abuse should be put in place. These can include limits on the size of transactions, stipulation of terms, and conditions on transactions or other aspects that regulators deem necessary.

If self-dealing is allowed, there must be open competitive procurement for all of it, overseen by regulators. Competitive procurement entails noticed proposals for procurement with specifications generally available and the outcome of bidding reviewed by regulators before contracts are awarded.

Regulators should have the authority to conduct ongoing oversight over the sale of competitive services to: ensure that they are in the public interest, protect competition, prevent provision of these services from jeopardizing ratepayers, and guarantee progress toward the goal of universal service.

Access to books and records must be allowed to facilitate regulatory oversight. This requires that federal and state regulators be given access to all books and records of both the affiliate and the operating company.

Sustained regulatory complaints of anti-competitive activity should be penalized by a freeze on all competitive activity, pending an audit of all activities and divestiture of activities directly related to the area of abuse.

In the new environment, pricing policies will have to be modified to allow competitors to utilize network functionalities on an unbundled basis. Rules must be developed to ensure that competitive services are not cross subsidized or priced in a fashion that unfairly disadvantages competitors. Rules must also be developed to ensure that captive ratepayers are not abused and receive the full benefits from the costs that they are bearing. Furthermore, rules must be developed for the purchase of bottleneck facilities and functionalities and public functions embodied in the public switched network.

While incumbent monopolists emphasize that the allocation of too many costs to competitive or discretionary services may price telephone companies out of the market, policymakers must also recognize that allocation of too few costs to competitive services may price competitors out of the market or place unfair burdens on captive ratepayers.

When incumbents are allowed to sell enhanced services, competitive tests must be established—not only for identical services, but also for close substitutes. Actual competition is the test for declaring a service competitive and subject to reduced regulation. Contestability as a substitute for competition is unacceptable. Actual competition should be measured by traditional market-share concepts. Rules for reclassifying services as non-competitive must be in place if the services do not prove to be competitive over the long term.

Pricing flexibility has little direct relationship to setting the rate of return as such. The revenue requirement is set independently of any individual rates. Pricing flexibility can be allowed, regardless of how the revenue requirement is set. Flexibility should be afforded only where there is actual competition and safeguards must be provided to both ratepayers and competitors.

The prices offered to unaffiliated companies should be considered a competitive standard for determining the bid price for self-dealing.

Chapter 12: Protecting the Public Interest in the Transition

Design and deployment of investment to be utilized by competitive services must be open to review by competitors and subject to regulatory oversight.

Network access services must be unbundled to allow potential competitors to capture as much value added as they desire. Non-discrimination in access requires cost-based tariffing of all services. Affiliates and telephone operating companies should not be permitted to engage in private negotiations. Provision of access services should be on a non-discriminatory, tariffed basis nor should the monopolist be allowed to be the first user of a service except after a reasonable period of wide public availability. Expedited procedures must be implemented to allow competitors to petition for the provision of specific services where a reasonable basis for such services exists.

The problem of the prevention of cross subsidies and the allocation of joint and common costs is crucial to managing the transition to competition. There is a massive basis for cross subsidy in the integrated firms. Integrated activities will commingle hardware (facilities) and software (expertise and resources) between regulated services, competitive services and services that are not regulated some at the state level and some at the federal level.

That there will be joint and common costs in the integrated companies is inevitable. Not only are facilities likely to be shared, but expertise to be shared would include personnel and software for routing, billing, and operations support systems—including traffic management, planning, and engineering to name just a few functions. Many of these managerial functions could be performed on a centralized basis. Indeed, we have witnessed a strong trend in the industry toward centralizing functions from many states in single locations. Initial and ongoing transactions between the regulated and unregulated components of the holding companies will abound.

As long as there are joint and common costs with some lines of business above the line and others below it, there will be an incentive to put costs above the line and profits below. Price cap regulation, which is touted as a panacea for all cross-subsidy problems, does not alter this incentive for services that are not subject to the cap. The incumbent monopolist does not need to recover all of the costs it misallocates from ratepayers to increase its profits. It may, in fact, increase its overall profits by shifting some costs to the monopoly in order to achieve a higher market share and a higher profit in the unregulated lines of business. Price caps do not eliminate this incentive.

Moreover, the problem of protecting the public and competitors goes far beyond the issue of cross subsidy. There are vast cost advantages that the franchise monopolist enjoys as a result of adding new businesses to its core monopoly. The existence of these cost advantages raises a fundamental question of whether stockholders or ratepayers have a claim on them and how they will impact on the competitive marketplace. If facilities have been ratebased and are being paid for by ratepayers, then ratepayers have a claim.

For example, the fiber optic trunks and loops being deployed by the Local Exchange Carriers (LECs) are vastly underutilized. That excess capacity, which will be used to provide video dialtone, is being paid for by ratepayers. However, some LECs have asserted that none of the costs of the fiber should be attributed to video dialtone. Ratepayers pay for the facilities used by video dialtone and stockholders reap the profits, since video dialtone revenues are considered competitive. Competitors of the LECs do not have access to such free facilities, since they lack a captive ratepayers. They are immediately placed at a competitive disadvantage. Perhaps the clearest statement of this approach to deploying an advanced network was made in internal BellSouth papers.

> This offers [1] the opportunity to cover the fixed costs of providing fiber to the home with [plain old telephone service] POTS revenue and selling [cable TV] CATV transport to overbuilder, entrenched CATV operator and pay service vendors (HBO, etc.) alike at probably market prices well in excess of incremental costs. At that time, [2] profit or rate-of-return regulation should have evolved to price regulation either by the current set of state and federal regulators or by the market itself. [3] This means BellSouth will be able to keep its CATV transport profits despite the relative low level of incremental cost required to provide the service.
>
> Having become "The Guy Who Got Fiber To the Home First," BellSouth's ubiquitous CATV transport will provide the "critical mass" necessary to support transport of the entire spectrum of BISDN services provided by the ESPs [Enhanced Service Providers]. Given the relative low incremental cost of "mining"

Chapter 12: Protecting the Public Interest in the Transition

more of fiber's huge bandwidth capacity to transport the wide variety of BISDN services and the pent-up demand signaled by the McKinsey study, BellSouth's profit potential appears good.[1]

In this case, the minimum cross-subsidy prevention standard does not adequately protect consumers, because they receive none of the benefits of the utilization of excess capacity for which they are paying. It does not adequately protect competitors because they are placed at a cost disadvantage that has its origins in the monopoly franchise, not in the competitive marketplace.

In the example given above, there may or may not be a cross subsidy, but the line is difficult to draw. If the cost of the excess capacity can be avoided, then there is a cross subsidy (telephone subscribers are paying more than the stand-alone costs of a well-engineered network).

There will be no cross subsidy if excess capacity cannot be avoided due to the fact that capital is lumpy and the excess capacity occurred in the pursuit of the least cost technology. Even if there is no cross subsidy, there is still a serious problem of economic coercion. Ratepayers receive no benefits from economies of scale and scope even though they bear the costs, and competitors are placed at a unfair disadvantage because they have no access to those economies. Cost and price analysis, which is crucial in preventing these problems, are discussed in the next section.

The principle of minimizing the burden of joint and common costs on captive ratepayers serves to protect both consumers and competitors.

The key concept is a "user pays" principle. All users of the advanced telecommunications network should pay for all functionalities that they use in reasonable proportion to the costs associated with those functionalities.

Floor prices (e.g., above long-run incremental costs) and ceiling prices (below stand-alone costs) should be identified to prevent cross subsidy and establish the range of acceptable prices. Subsidy-free pricing is the economic efficiency standard that must be met.

---

[1] Memo on "CATV Transport: Catalyst for BISDN" from R.T. Burns, Assistant Vice President, to N. Baker, Senior Vice President, BellSouth Services, June 14, 1988, p. 10.

Subsidy free pricing only establishes a range of prices that are reasonable. Where joint a common costs are large, this range of prices will be extremely wide. Subsidy free pricing is the necessary condition for economic efficiency.

Where flexibility in pricing exists, pricing methodologies should minimize prices charged for basic services to captive ratepayers. When there are unallocable common and joint costs in enterprises selling a combination of competitive and monopoly services, the contribution from competitive services should be maximized.

Because captive ratepayers have no alternatives, regulatory mechanisms must protect them from excessive burdens. Minimizing the burden on ratepayers and maximizing the contribution of competitive services also protects competitors from unfair competition because competitors do not have access to a captive monopoly-core business to absorb costs.

Specific, predictable price rules—e.g., equal mark-ups above direct costs or equal mark-downs below stand alone costs—must be applied to ensure that competitors are not placed at a disadvantage and that consumers are compensated for the costs of facilities used to provide competitive services.

The user-pays principle, combined with the principle of minimization of burden on basic services, is the key to affordable universal service. In light of the immense capabilities of the network, these two principles will provide a sound basis for affordable service.

The burden of joint and common costs placed on basic access should be minimized for a number of reasons. As a matter of social policy and in recognition of the economic value of having more people on the network (i.e., the network externality) basic service should be a low mark-up service. Because captive ratepayers have no alternatives, regulatory mechanisms must protect them from excessive burdens. Minimizing the burdens on ratepayers and maximizing the contribution of competitive services also protects competitors from unfair competition because competitors do not have access to a captive monopoly-core business to absorb costs.

As enterprises become involved in a mixture of monopoly and competitive services, additional risks may be incurred and benefits may be conferred on the monopolist. Regulatory mechanisms to insulate basic

Chapter 12: Protecting the Public Interest in the Transition

service ratepayers from the risk of competitive enterprises must be established. Increases in the cost of capital caused by those enterprises must fall on competitive businesses.

There are a variety of economic advantages gained by the incumbent monopolist. These, too, provide an economic basis for lowering basic access rates. Many of the activities into which the incumbents would like to move and have moved benefit in tangible and intangible ways from the fact that they are extensions of the franchise monopoly. The people who grant the franchise have a right to share in the economic benefits that the monopoly creates.

Unregulated subsidiaries should not be allowed to achieve excessive rates of return because they are an extension of the franchise monopoly. Revenue streams resulting from readily identifiable company monopolies (e.g., the Yellow Pages) should be carried above the line for regulatory purposes. Cost-reducing advantages for competitive services that flow from the monopoly franchise (e.g., new subscriber lists) should be recognized by fees paid to monopoly services. The value of intangible benefits (e.g., good will) should be estimated and paid for to lower the cost of monopoly services. These are not mere paper transactions, because the economic reality underlying the network ensures that the cost advantages enjoyed by the local franchise monopolist can be exploited in reality.

This approach to pricing is exactly the opposite of the Ramsey pricing rule (it is the inverse of the inverse elasticity rule). Ramsey pricing has been consistently rejected by regulators, and this is crucial for an efficient and fair transition to competition. The weaknesses of Ramsey pricing are clear in the empirical literature.

◆ As a theoretical proposition, the Ramsey pricing rule rests on an extreme and extensive set of assumptions that have virtually no chance of being met in reality.

◆ As an empirical matter, the Ramsey pricing rule is intractable, since it requires data on demand elasticities that are not available.

◆ As a matter of public policy, the failure to meet theoretical assumptions and the weakness of the data mean that the Ramsey pricing rule guarantees neither efficiency nor procompetitive outcomes.

◆ Morally and legally, the pricing rule is questionable because it places the heaviest burden on captive ratepayers.

Some have argued that rate rebalancing and stranded investment created by competition potentially creates a pricing nightmare for captive consumers. That is not necessarily the case.

First, competitive pressures may or may not develop. They may or may not require increases in local rates. They may first result in forcing incumbent monopolist profits back down to reasonable levels. They may cause the companies to reconsider their cost allocation approaches. They may cause the companies to be more careful in their investment decisions.

Second, if technology has rendered the incumbent monopolists' approach to provision of basic service uneconomic it may cause the companies to write down their assets, as firms facing competition frequently do. The risk premiums enjoyed by incumbents over decades should not be ignored in this analysis. Companies that insist on recovering all capital costs, even when they have been rendered obsolete, are asking for a risk-free investment. The possibility of having to write off investment has already been compensated for in the risk premium allowed and earned by most utilities.

Only after the rigors of real competition have exposed the true economics of basic service should we consider subsidies necessary to prevent local rate increases and preserve universal service. This is the topic of the next section.

Having argued that shareholders should be forced to shoulder the burden of stranded investment, where such investment has been rendered obsolete by technological progress or uneconomic by competition, we must also recognize that there are certain public functions that may not be "economic" for individual investors in a competitive market place. These must be compensated to ensure their delivery.

In order to achieve the conditions required to open the local exchanges to competition while preserving the essential function of the network and protecting captive ratepayer, regulators must dissect the network into each of its constituent elements, cost each one separately, and ensure that competitors bear a fair share of these costs. Revenues must be raised to cover these costs. There are a variety of approaches possible.

Chapter 12: Protecting the Public Interest in the Transition

First, for discrete services that competitors are likely to want but not to want to pay for, tariffs can be established. For example, it makes little sense to have multiple suppliers of 911 service. Competitors are almost certain to want to have access to 911 service, however. Competitors, both incumbents and new entrants, should be charged at least a cost-based rate for use or access to such services, and the revenues used to lower basic service rates.

Second, the obligation to service must be maintained. It need not fall on a specific franchise holder, however; develop other mechanisms that ensure service but are more neutral. Several alternative approaches could be taken here.

One such approach is to identify customers who need support to maintain service at affordable rates, either because the cost of service is too high (high-cost funds) or because their income is too low (lifeline rates). Funds to maintain service for these customers can be raised by across-the-board taxes or fees on all service providers that are then transferred to those companies that serve these customers.

A second approach is to identify exchanges that are deemed unprofitable by the incumbent. Alternative providers are invited to bid to serve these exchanges. The provider with the lowest subsidy necessary to serve the exchange wins the bid.

Conclusion

The regulatory agenda for the transition to competition requires both structural and accounting protections. To promote competition, regulators must unbundle network functionalities and prevent economies of scale and scope from being used to undermine competition. Structural separations are necessary to facilitate these protections, but cost and price analysis is central to effective regulation.

All users of the network must pay for all the functionalities they utilize in proportion to the nature of the demand they place on the network.

Incremental and stand-alone costs for all services should be calculated to identify the range of subsidy-free prices. Prices that fall in this range meet the basic condition of economic efficiency.

The range between incremental and stand-alone costs will be large because joint and common costs are large in network industry. The recovery of these costs should not place a heavy burden on captive basic service ratepayers for economic reasons (to prevent incumbent monopolists from disadvantaging competitors) and public policy reasons (to promote universal service). Consequently, rules such as Ramsey price rules, which run counter to these proconsumer and procompetitive outcomes, should be rejected.

Imputation of costs for functionalities used by competitors is crucial to protecting captive ratepayers and promoting competition.

Excess profits or inefficiencies in production should not be recovered in the prices charged to ratepayers. This framework establishes an empirically manageable, balanced approach to pricing. It sets economic efficiency criteria, without pursuing economic efficiency to extreme, burdensome, and often unachievable ends. With basic conditions of efficiency met, it blends in public policy goals by protecting captive ratepayers from bearing an unfair share of joint and common costs and protecting competitors from economic coercion by incumbents who allocate all the economies of joint and common production to excess profits and/or anti-competitive prices.

Part 5

The Electric Industry in Transition

INTRODUCTION

Transition Costs with Increasing Competition

Michael J. Kelleher
Director, Economic Research and Forecasting
Niagara Mohawk Power Corporation

In the United States airlines, natural gas, railroads, telecommunications, and trucking have undergone relatively recent deregulation or divestiture. In each case the competitive marketplace produced innovative new products and services, lower prices, cost reductions, and pricing structures that had not been seen in the regulated industries. The change to competitive markets has also brought bankruptcy, mergers, and acquisitions as firms evolved to meet the rigors of the competitive marketplace. The change brought about by competitive markets is rapid and persistent.

Regulated electric rates have risen to levels that have provided electric consumers with alternatives and existing competitors with opportunities. Rising electric rates, combined with falling natural gas and oil prices, have driven existing electric customers to switch fuels. The same forces have spurred the increase in cogeneration and self generation by our customers. The suppliers of oil, gas, and small generation technologies have responded vigorously and rapidly to the market conditions, in sharp contrast to the regulated utilities' response.

The pressure from retail competition has been felt increasingly by regulated electric utilities in recent years. The primary determinant of such competition has been economic pressure, that is, the difference between the incremental or marginal costs of providing electric service and regulated prices. Without retail wheeling the primary competitive threats to electric utilities are limited to fuel switching, self generation, cogeneration, and the movement of ratepayers to other geographic areas. While certainly not trivial, the economics that underlie these competitive pressures are far less troubling than those of retail wheeling. The economic discounts promised by proponents of retail wheeling have provided a powerful signal for customers with which to try to obtain access to low-cost power. Authors in other sections have focused on the inevitability and desirability of retail wheeling or direct access. The three papers in this section focus on the cost of making the transition to competition in the electric industry.

The three papers in this section differ somewhat. Dr. Theresa Flaim describes a variety of commitments that utilities have made under current regulations, and estimates the value of stranded commitments. In addition, Flaim describes the advantages and disadvantages of a variety of potential methods to deal with stranded commitments. She discusses the impact regulatory policy can have on the level of stranded costs in the industry, and focuses particular attention on the question of the timing of retail competition.

The change to a competitive market will also likely involve significant restructuring in the industry. Dr. Charles Stalon describes the need to vertically de-integrate the electric industry. If the generation, transmission, and distribution functions of electric utilities are not separated, Stalon argues, developing a competitive market will be more difficult. Stalon also describes distribution, transmission, and grid control functions as natural monopolies that will require continued, but improved, regulation. There has been debate about the form of industry restructuring and possible vertical de-integration. Stakeholders in competing generation entities think vertical de-integration is necessary to insure their equal access to the transmission system and retail customers.

Dr. Edward Kahn describes the responsibilities of government, particularly the federal government, in the evolution of the electric industry to a more competitive form. Kahn argues that the federal government was a strong proponent of the formation of nuclear generation. As such, and because there are no private sector solutions for nuclear stranded costs, the federal government bears a responsibility in solving issues of stranded costs related to nuclear generation. Kahn also describes the necessary oversight by government to ensure safety and help solve the future challenges of nuclear decommissioning.

The three papers together describe many of the challenging details involved in developing transition policy for a move to a competitive electric industry. The authors also pose potential solutions that offer policymakers alternatives to consider in guiding the transition.

CHAPTER 13

Methods for Dealing with Transition Costs for the Electric Utility Industry

Theresa A. Flaim, Ph.D.[1]
Vice President, Corporate Strategic Planning
Niagara Mohawk Power Corporation

An important problem associated with rapidly increasing competition in retail electricity markets is that it would create transition costs. Transition costs exist because of a series of commitments that were made by customers, regulators, lawmakers, and utilities that competition would make difficult to keep unless specific provisions are made to account for them. Commitments made under the current regulatory system might become stranded if electricity markets are competitively restructured and the resulting market price were too low to allow for recovery of the costs associated with those obligations.

There are three kinds of these commitments.

◆ Stranded assets are investments (primarily in generation plant) that utilities prudently made to keep their commitment to provide service to all customers at high levels of reliability in return for the regulatory commitment that these expenses would be recovered over time.

◆ Stranded liabilities are primarily contracts with unregulated generators whose price terms and conditions were dictated, in large part, by state law and regulatory policy.

◆ Regulatory assets are primarily deferred expenses such as previously flowed-through tax benefits that appear as assets on the balance sheet in return for the regulatory promise that the utility will be allowed to recover them in the future.

---

[1] This paper is in large part excerpted from a Niagara Mohawk Power Corporation Report "The Impacts of Emerging Competition in the Electric Utility Industry," issued on April 7, 1994. Nevertheless, the views expressed in this article are those of the author, not necessarily those of Niagara Mohawk.

◆ Stranded social programs include a variety of social programs that utilities have implemented in the past through their monopoly power of taxation. Such programs include tax collection, environmental compliance beyond that required of firms in competitive industries, the universal obligation to serve (often at rates below cost), cross-subsidized pricing of all services (including demand-side management), special programs for low-income customers, and so forth. The potential size of transition costs for the investor-owned electric utility industry cannot be overstated. Estimates of potential stranded asset exposure in the United States electric utility industry as a whole range as high as 150 to 200 billion dollars, compared to a total shareholder equity of 180 billion dollars. Thus stranded commitments for utilities are potentially more than ten times the total transition costs estimated for the natural gas industry. Because of their potential enormity, a fair and equitable plan must be developed for dealing with them.

This paper will discuss the advantages and disadvantages of the various methods available for dealing with transition costs. Before proceeding with the analysis of alternatives, several background comments are in order. First, the analysis of options set forth in this paper is based on four underlying assumptions that are themselves subject to vigorous debate.

◆ Competition would result in real cost savings in the utility industry.

◆ Increased competition at the wholesale level alone is unlikely to achieve the full benefits of more vigorous competition at the retail level.

◆ Global competition is the underlying driver of increased competition in the electric utility industry. Increasing competition in electricity markets is therefore inevitable.

◆ Interregional competition and economic development are great concerns in most states at present, particularly in New York State.

With these underlying premises, the basic methods for dealing with transition costs can be evaluated. They are as follows:

◆ Use time to allocate transition costs. One extreme would be to open markets immediately and let the financial chips fall where they may. The other extreme would be to slow the pace of competition as long as possible.

Chapter 13: Methods for Dealing with Transition Costs

- Reduce utility costs and use the savings to offset accelerated recovery of stranded assets and liabilities, or to fund social programs.

- Redistribute existing utility costs among customers, competitors, taxpayers, and shareholders using appropriate fees and changes in rates.

These three basic approaches are not mutually exclusive. In fact, all could and will probably have to be adopted in combination where transition costs are large. Before addressing each one in turn, however, it will be helpful to address the question of utility asset writedowns. When addressing the question of utility asset writedowns, it is important to remember that asset writedowns for firms subject to cost-based regulation have far different consequences than for firms in competitive industries. For firms subject to cost-based regulation, a writedown means that shareholders are permanently giving up their rights to recovery of that investment. Thus, it is a permanent reduction in shareholder value. For firms in competitive industries, a writedown has no effect on the prices they are allowed to charge, and therefore, no impact on shareholder value.

In discussions about the impact of emerging competition in the electric utility industry, the debate often focuses around the following question: "How large a writedown should utilities be forced to take?" This is the wrong question and, in fact, presumes the answer to the real question that needs to be addressed: "Who should pay for the costs of making the transition to more competitive markets?" Obviously, this question is fundamentally one of law, equity, and fairness.

Use Time to Allocate Transition Costs

Transition costs are directly related to the timing of the transition. For example, if the electric industry had been deregulated in 1965, market prices would likely have exceeded regulated prices (at least initially), and there would have been excess profits in the short run rather than stranded commitments. Looking ahead, full deregulation in the next five years would create very large stranded commitments due to the excess capacity situation in the Northeast. If policymakers believe that compensation for stranded commitments is appropriate, then a very attractive practical solution might be to slow the pace of competition to allow time for the New York utility system to grow out of its current capacity surplus and mitigate the need to make difficult reallocations of transition costs. To make such a decision, one would have to weigh the potential benefits of competition in the near term against the huge transition costs that would be created if the transition to competition were made abruptly.

Time is obviously a continuum. However, it is useful to examine the two extremes of a spectrum. At one end is the option of opening markets immediately and at the other slowing the pace of competition as long as possible.

Open Markets Immediately and Let the Financial Chips Fall Where They May

Advocates of this approach argue that opening the markets as quickly as possible would allow the forces of competition to work immediately, put downward pressure on utility prices, and minimize transition costs for ratepayers. While acknowledging that such a move would have severe financial consequences for utilities—to the point of cascading bankruptcies in certain parts of the country—advocates of this approach simply argue that "eventually the market will sort it all out."

This approach also has important disadvantages. First, it would have catastrophic financial consequences for the utilities, who are not responsible for the current excess capacity situation and the resulting gap between current prices and the price that would exist if markets were opened immediately. Second, if any attempts are made to open up markets without a plan to deal equitably and fairly with transition costs, utility managers would have no alternative but to oppose such initiatives as a way to protect shareholders' investment.

Third, under severe pressure to reduce retail prices a utility's ability to carry out social programs would be greatly impaired. Last, reliability of service would likely deteriorate as utilities struggled to reduce costs and found it more difficult to raise capital in their weakened financial condition.

Slow the Pace of Competition as Long as Possible

Slowing the pace of competition would help deal with transition costs because higher prices could be maintained long enough to recover commitments that would otherwise be stranded. It would also avoid the need to reallocate utility costs abruptly and would allow utilities time to prepare for competition.

However, delaying competition indefinitely is not feasible since some forms of competition are already here today. In addition, to do so will delay any cost savings that result from increased competition. Another concern is that incentives to become more efficient may not be as strong as they would be if electricity markets were fully competitive. Finally, if

Chapter 13: Methods for Dealing with Transition Costs

the utilities themselves are forced to seek a delay, this strategy can require time and resources that might be better used preparing for competition.

Despite the problems associated with the two extremes on the time spectrum, state and federal regulators and legislators have a great deal of control over how competition is allowed to unfold in retail markets.

If a chaotic transition is to be avoided, a public policy decision needs to be made about the extent to which competition should be allowed to increase in electric retail markets, and then a plan for an orderly transition needs to be developed. That plan will need to include a timetable for allowing various markets to be opened to competitive alternatives, assuming that such is the outcome desired. The appropriate timetable for the plan will depend on how large the stranded commitments are: the larger the stranded commitments, the more intractable are the alternatives for dealing with them.

Reduce Utility Costs and Use the Savings to Accelerate Recovery of Stranded Commitments

Two categories of utility costs need to be addressed: direct costs (including capital and expenses) and incumbent burdens. A utility's direct costs are those costs over which the utility has substantial control. They include departmental expenses, capital spending, fuel, and so on. Niagara Mohawk is taking aggressive steps to reduce its direct costs, including staff reductions of 1,550 (planned for 1995) and a reduction in planned capital spending of roughly 120 million dollars per year.

Incumbent burdens are things that utilities are required to do by law, regulation, or custom that unregulated competitors generally are not required to do. The incumbent burdens that have a major impact on Niagara Mohawk's total cost structure include excess payments to unregulated generators, higher taxes, and the obligation to serve all customers—often at rates below cost.[2] If utilities are to reduce costs to become competitive, then both direct costs and incumbent burdens must be addressed.

Incumbent burdens account for a substantial proportion of Niagara Mohawk's total cost structure. Figure 6-1 shows the allocation of projected 1995 electric revenues to functional accounts. Taxes and payments to unregulated generators together account for 46 percent of

---

[2] Theresa A. Flaim, direct testimony in PSC Cases 94-E-0098, 94-E-0099, and 94-G-0100, February 4, 1994, pp. 19–26.

The Electric Industry in Transition

Unregulated Generators 28.7%
2.7 cents

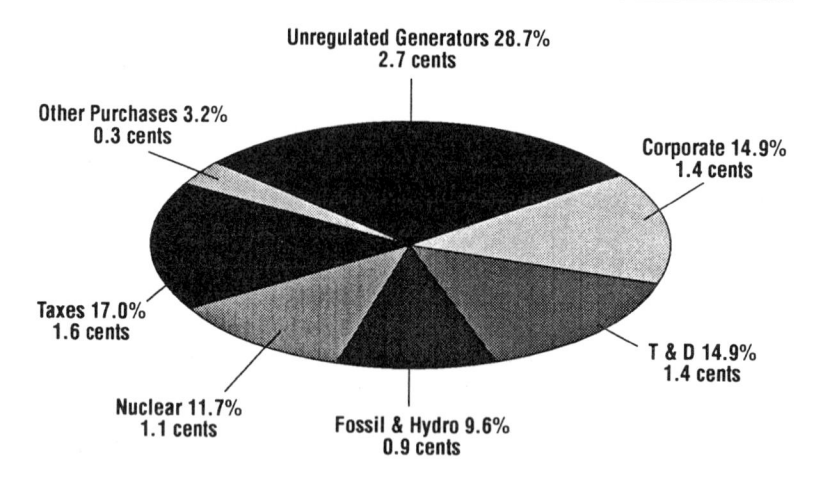

Other Purchases 3.2%
0.3 cents

Corporate 14.9%
1.4 cents

Taxes 17.0%
1.6 cents

T & D 14.9%
1.4 cents

Nuclear 11.7%
1.1 cents

Fossil & Hydro 9.6%
0.9 cents

Figure 14-1. Allocation of Projected 1995 Electric Revenues by Functional Account.

the projected revenue. Although the separate impacts of other incumbent burdens are small relative to that of taxes and unregulated generator payments, the collective impact of the following is also important: cross-subsidized pricing of all services (including demand-side management), low-income assistance programs, state-mandated minimum hook-up requirements combined with rate structures that are inadequate to recover associated costs, economic development subsidies, and asymmetrical environmental liabilities. The Company believes that incumbent burdens limit its ability to control more than half of its total costs.

For example, as part of its attempts to lower its tax burden, Niagara Mohawk has challenged the City of Dunkirk's property tax assessment of its generating facility there. In response, the city trebled the assessment from 70 million dollars to 196 million dollars in one tax fiscal year. Niagara Mohawk has appealed the assessment. The matter is now in the courts and will likely take months (if not years) to resolve.

Similarly, unregulated generator costs are extremely difficult to affect. Reducing these costs to any significant degree requires contract renegotiations, and when those renegotiations prove futile, litigation is involved. An additional problem is that Niagara Mohawk's ability to enforce the contracts it has signed in the past—particularly with respect to pricing and curtailment provisions—has been constrained in many cases by laws and regulatory policy.

There are two major advantages associated with the strategy of reducing utility costs and using the savings to accelerate recovery of stranded assets and liabilities or to fund social programs. First, it allows recovery of those costs while minimizing rate increases for core customers. Second, increasing competition in electricity markets will put continuous downward pressure on prices and, ultimately, costs will have to be reduced to respond to that pressure. Thus a strategy that reduces costs as a way to fund stranded commitments is consistent with where electricity markets are going.

There are two disadvantages associated with this strategy. One disadvantage is that even though this strategy would minimize rate increases, it would delay any price decreases associated with cost reductions until after the stranded commitments are recovered. A second disadvantage is that—to the extent that incumbent burdens are reduced by reducing social program obligations—the associated programs and their benefits would also be reduced. On balance, however, this strategy is appropriate and reasonable and should be pursued to the maximum extent practicable.

Redistribute Utility Costs

When transition costs are large, utility cost savings alone are unlikely to recover stranded commitments, unless competitive forces are minimal, for many years. Thus some redistribution of stranded commitment costs will undoubtedly be necessary as part of an equitable plan for dealing with the transition to more competitive markets.

Two questions related to redistributing costs are relevant. First, how should transition costs be distributed among core customers, non-core customers, taxpayers, competitors, and shareholders? Second, what are the mechanisms for collecting costs from those various groups? The problems associated with assigning transition costs to shareholders were discussed previously in the context of asset writedowns and in the context of using time to allocate transition costs. This section will focus on the options for redistributing costs among core and non-core customers.

Should Core Customers Pay Transition Costs?

Core customers are customers with virtually no competitive alternative to utility-supplied electricity. Many—but not all—core customers are small in size. An advantage of redistributing utility costs to smaller customers is that a redistribution would actually reflect the underlying costs of serving those customers more accurately. Because of the industry's long tradition of collecting fixed costs volumetrically, large customers have subsidized small customers both within and between classes for years. At a minimum, redistributing costs so that small customers are paying prices that reflect the fully distributed costs of supplying them is clearly appropriate.

If stranded commitments are large enough (which will be a function of how quickly competitive forces are allowed to extend into retail markets), some additional redistribution of utility costs to core customers might be necessary to preserve the financial viability of the utility and the reliability of the system. Finally, to the extent that price discounting promotes economic development core customers benefit from the increased economic activity as well.

There are two disadvantages to redistributing utility costs to core customers. One is that it is politically unpopular to raise rates for small customers faster than for other customers. (It is not by accident that small customers have been subsidized for so many years.) Also, there are competitive limits to how much rates, even for core customers, can be increased. For example, Niagara Mohawk is losing roughly 5,000 residential electric space-heating customers a year. Any redistribution of stranded commitments must reflect the realities of the competitive market.

Should Non-core Customers Pay Transition Costs?

Allocating some transition costs to non-core customers is certainly fair and appropriate, since transition costs represent commitments that were made on behalf of all customers and since such a strategy would help fund transition costs from the customers who are benefitting most from increased competition. The difficulty, of course, is that it may not be feasible to collect them from customers with competitive alternatives. For example, customers who reduce their consumption can easily avoid transition costs that are charged volumetrically. Similarly, it would be difficult to collect transition costs from customers who relocate or shift load to another region. Nevertheless, this strategy should be pursued to the maximum extent practicable.

Chapter 13: Methods for Dealing with Transition Costs

Should Some Transition Costs be Shifted to Taxpayers?

Certain incumbent burdens that were established as a matter of convenience, but which have little to do with electricity service *per se*, should probably be shifted to taxpayers. Obvious programs in this category are tax collection and certain forms of low-income assistance.

Methods for Redistributing and Collecting Costs

There are four methods for redistributing and collecting transition costs: (1) volumetric surcharges, (2) exit and access fees, (3) shifting from cost-based to market-based valuation of all assets, and (4) accelerating the depreciation of potentially stranded assets.

Volumetric surcharges. Once calculated, transition costs can be charged on the basis of usage. The difficulty with this mechanism is that it charges larger users more than smaller users. To the extent that large users have more competitive alternatives than small users, this approach could encourage uneconomic bypass.

Exit and access fees. Exit fees provide a method for collecting transition costs from the customers who directly benefit from competition. For example, exit fees are probably appropriate today for new municipalization proposals. Customers should not be allowed to form new municipalities simply to avoid the costs of commitments that were undertaken on their behalf. To allow that to happen would merely shift those costs to the remaining customers and shareholders. If a policy on exit fees for municipalization is established now, customers will be able to evaluate such proposals on the basis of their true economic merits and not simply on the basis of their ability to avoid sunk costs.

Fixed-cost access fees will allow transition costs to be collected from all customers, regardless of their usage. Thus they would not encourage some forms of uneconomic bypass.

There are several disadvantages to exit and access fees that would need to be addressed in their design. First, if the exit fee associated with onsite generation were large enough, it would encourage customers to bypass the system for backup service as well. Exit fees should not be set so high that they encourage uneconomic bypass of the utility. Second, we would have to insure that access fees do not inappropriately discourage new load. Third, exit and access fees will not work for load loss due to all forms of competition, such as interregional competition and plant relocations. Such fees would have to be designed to be consistent with economic development objectives.

Shifting from cost-based to market-based valuation methods. Writing up transmission and distribution assets should be considered if generation assets are to be written down. A writedown of generation assets is, in effect, a shift from a cost-based valuation to a market-based valuation. If that step is contemplated, then the writedown should be of all assets, not only those whose market-based valuation might be less than their cost-based valuation. Replacement cost would be a reasonable basis for valuing the market value of transmission and distribution assets.

If stranded commitments are larger than the amount of revenue associated with the replacement-cost valuation of transmission and distribution assets, they could be written up to the amount necessary to recover stranded commitments in total. The advantage of this approach is that it could separate the recovery of stranded commitments from competitive pricing in generation if that is the desired public policy goal.

To insure recovery of costs, this proposal will have to be designed in conjunction with exit or access fees. Otherwise, costs would be reassigned according to traditional cost-of-service methods, and customers might avoid payment of the costs through uneconomic bypass of the system (especially to the extent that such costs would be allocated volumetrically). Customers on the border of two service territories might avoid payment of a higher wheeling charge, for example, by building a line to a neighboring system. The proposal might also reduce the competitiveness of Niagara Mohawk's wheeling services in wholesale markets, but that problem is probably quite small relative to wheeling's potentially significant benefits.

Finally, if a change in asset valuation affects rates for wholesale transmission services, the FERC would also have to approve the change.

Accelerating the depreciation of assets that might become uneconomic under competitive market pricing should be considered, but only if it can be accomplished without encouraging uneconomic bypass. Thus this strategy probably makes the most sense if it can be offset by cost savings elsewhere.

A Road Map for the Transition

The transition from an electricity industry dominated by regulated monopolies to one that is characterized by much greater competition in retail markets poses significant challenges for utilities and regulators alike. A roadmap for the transition should contain the elements listed in the following paragraphs.

Utilities Must Aggressively Manage Costs

Competition will put continuous downward pressure on retail prices. Utilities will, therefore, be challenged to find every opportunity to reduce costs while maintaining high levels of service quality and reliability.

Incumbent Burdens Must Be Reduced, Shifted, or Funded through a Tax on All Competitors

Because competition will continue to erode monopoly control of markets, it will also erode utilities' power to tax customers to fund social programs. Thus competition will mean that the incumbent burdens on utilities will have to be reduced, funded from a shrinking base of core customers, shifted to taxpayers, or funded through a tax on all competitors.

As a practical matter, competition will mean that at least some portion of incumbent burdens will have to be reduced. In selecting those utility mandates that should be preserved (albeit at a reduced or modified level), it makes sense to select those that are directly related to the utility services and reduce or shift those that are unrelated to electricity *per se*. Environmental, energy efficiency, and research and development programs, as well as programs aimed at maintaining universal service (such as low-income programs), are the most directly related to utility services.

Using utilities as tax collectors was administratively convenient for government in the past, but it is increasingly unsustainable and is not directly related to electric services. Thus the gross receipts tax, in particular, should be phased out and the burden shifted to other taxpayers.

Traditional Regulation Needs to be Modified to Be More Consistent with Competitive Markets

For reasons that are documented more fully in testimony elsewhere[3], traditional rate-of-return regulation is inconsistent with competitive markets in serious ways. Briefly summarized, traditional regulation has

---

[3] Theresa A. Flaim, direct testimony, *op cit.*

four problems. First, it provides only weak incentives to minimize costs and prices. Second, incumbent burdens severely constrain a utility's flexibility and its ability to reduce costs. Third, regulatory processes are too slow and burdensome, given how rapidly conditions in competitive markets can change. Finally, traditional pricing of utility services has resulted in prices that are inefficient and unsustainable under increasing competition. If the electric utility industry is to move in directions that allow and encourage increased competition in electric retail markets, the current regulatory system will have to be modified so that it does not handicap the utilities who are now being asked to compete.

Niagara Mohawk's filing for rate year 1995 proposes changes in regulation that are more consistent with emerging competition. Its key features are as follows:

- A five-year price caps plan offering better cost efficiency incentives and the opportunity to identify and pursue them

- Elimination of the Niagara Mohawk Electric Revenue Adjustment Mechanism (NERAM) providing strong financial incentives to minimize uneconomic bypass

- Reduction of cross subsidies

- Greater flexibility in pricing competitively in contestable markets

- Protection of core customers through price caps

- Acceptance by shareholders of the risk of returns up to 300 basis points below the allowed rate of return on equity in exchange for the opportunity to make greater productivity improvements and price competitively in contested markets

A Policy on Exit Fees for New Municipalization and Other Forms of Bypass

A policy on exit fees for new municipalization and other forms of retail bypass is needed now. Without such a policy, new bypass proposals, based on the apparent opportunity for some customers to avoid paying for stranded commitments and shift the burden to other customers and shareholders, can proliferate. Moreover, it is likely to be easier to establish such a policy before there is a proliferation of specific bypass proposals. At a minimum, the administrative burden of dealing with what could become a large number of cases will make it more difficult to concentrate on the development of a rational, generic policy.

Two forms of bypass are of concern. First is that unregulated generators can potentially serve retail customers. The second is municipalization. The Sithe/AlCan/Beloit case illustrates the first problem. Without a policy on exit fees or other recovery of stranded commitments, such cases could proliferate.

Far more troubling, however, is the potential for a mass movement toward new municipalization. Such a wave could happen after the FERC issues regulations governing wholesale open access. Independent wholesale power brokers are already forming in anticipation of having greater access to markets. Given the current glut of supply in the Northeast and Canada and the dearth of existing wholesale markets, it is inevitable that such brokers will begin to market new municipalization proposals as ways to market surplus power.

The greatest appeal of both forms of retail bypass will be in the potential they offer for gaining access to generation at current severely depressed wholesale spot market prices and for avoiding sunk costs and other incumbent burdens. If customers considering bypass alternatives know in advance that they will have to pay their fair share of commitments that were made on their behalf, they will be able to evaluate such proposals based on their true economic merits.

If Retail Competition is to Increase Beyond Current Levels, a Plan for an Orderly Transition Is Needed

As a practical matter, regulators and legislators cannot stop the competitive forces that have already been unleashed on the electric utility industry, but they have a great deal of control over their pace and direction in the future. More importantly, regulators and legislators also have control over utility costs in New York State. In Niagara Mohawk's case, incumbent burdens account for far too large a proportion of total costs to justify the belief that reducing costs to competitive levels is achievable by the company acting alone.

Indeed, negative reaction on the part of special interest groups to the company's attempts to reduce these costs has been striking. One notable example is the City of Dunkirk's response to the company's challenge of the property tax assessment mentioned above. Another is the vehemence

The Electric Industry in Transition

with which some demand-side management (DSM) advocates, calling the changes "shocking"[4], have reacted to the company's attempts to restructure its DSM programs so that they are consistent with where electric markets are going. Utilities caught between mounting competition and the strident reactions of special interest groups find regulatory indecision about the direction in which they should proceed discouraging.

The importance of having an orderly plan that deals fairly with transition costs cannot be overemphasized. First, it would minimize the need to reallocate costs abruptly and dramatically to recover stranded commitments. Second, having an orderly plan would allow customers to make better energy-related investment decisions, because they would have a clearer idea of how their options might change in the future. Finally, it would allow utilities to focus on preparing for competition. Without such a plan, utilities might be forced to oppose attempts to open up markets, when those resources would be better spent preparing for competition to the ultimate benefit of customers and shareholders alike.

An optimal transition plan should:

♦ determine the extent to which it is desirable as a matter of public policy to increase retail competition in electricity markets significantly

♦ estimate the transition costs associated with achieving that level of competition over time

♦ develop a timetable for achieving that level of retail competition

♦ develop a method for fairly and equitably allocating transition costs

If the electric utility industry is going to make the transition to more competitive markets, all of those involved need to decide as a matter of public policy how the industry is going to get there and then get on with doing the things that will have to be done to make that transition. Niagara Mohawk looks forward to collaborating with policy makers, customers, and other interested parties to develop a plan that will, when implemented, help to insure a fair and equitable passage to increased competition in the electric utility industry.

---

[4] "Niagara Mohawk Would Drop Decoupling, Slash DSM, Under Five-Year Rate Plan," *Demand-Side Report*, March 3, 1994, p.8.

CHAPTER 14

Stranded Investments Costs: Desirable and Less Desirable Solutions

Dr. Charles G. Stalon
Consultant

All of us are paying a price today for living under a system of monopoly regulation that is not in tune with modern marketplace realities.[1]

Forecasting failures are a principal source of the electric industry's investments and commitments that we now believe to be "uneconomic."[2] Paradoxically, when stranded investments are defined as the difference

---

[1] Carlos A. Riva, "The 1994 Electric Executives Forum," *Public Utilities Fortnightly*.

[2] Two views of the "stranded investment" or "stranded commitment" problem deserve recognition. One view emphasizes that intensifying competition creates stranded investments. The second view emphasizes that intensifying competition only reveals stranded investments. The first view is illustrated by the comment of Bernard M. Fox in "The 1994 Electric Executives' Forum," *Public Utilities Fortnightly*, Vol. 132, No. 11, June 1, 1994, p. 39.
> Stranded investment could be created if current retail customers bypass their local utility system and thereby bypass paying for those costs incurred to serve them.

The second view is illustrated by the comment of Jeffrey K. Skilling, p.47, from the same source.
> [Stranded investments] are not new incremental cost brought about by increasing competition, but instead costs already being borne by utility ratepayers....The issue of stranded investments has surfaced because a more market-oriented approach to delivery of electricity has highlighted these inefficiencies."

The view supported in this paper is that there are important elements of truth in both views, and all defensible allocations of the costs of stranded investments will call upon both views for their justification. Restructuring does not mean that exactly the same set of services will be offered by the new industry that was offered by the old one. Investments (in some types of DSM mechanisms in the homes of customers, for one example) that depend on strong monopoly powers may not be recoverable in the new industry. That fact is not sufficient to justify an assertion that such investments are not appropriate, or were not "efficient," for a regulated monopoly.

between the book values of generating assets and contract rights and the market value of such assets, estimates of the magnitude of these stranded costs rely on still more forecasts of demands and fuel prices over the expected life of the assets. This fact ought to be a source of concern because there is no reason to believe that the art of forecasting has improved noticeably in the last two decades.

The Forecasting Problem for Electric Utilities

The capital intensity of electric industry assets and the long expected life of such assets makes it necessary to use forecasts to evaluate and to justify investment decisions. The reality, however, is that all long-term economic forecasts that purport to convey useful point estimates, in contradistinction to astrological forecasts, will be wrong. Serious economic forecasting starts with a recognition that the most important part of a point forecast is the variance of the estimate, and that variance is merely another point estimate with a variance that is another point estimate, etc. This conclusion holds whether the forecast is used to calculate the generation capacity needed or to estimate the magnitude of stranded investments.

Since these elementary economic truisms have been recognized by business and economic planners since the mists of early history, one question of interest before us is why the electric industry has made so many large "uneconomic" investments in the last quarter century.[3] I suggest that two intertwined mistakes have been made repeatedly; they were made in the early 1960s when generating capacity needs were forecasted and they were made again in the 1970s and 1980s when the "need" for the services of NUGs was forecasted. I stress this because I fear a repeat of these mistakes when forecasting the content and magnitude of stranded investments.

These mistakes are supported by a tendency of regulatory decision processes to generate certainty out of uncertainty. These processes tend to convey faulty information to all who depend on them for critical information. The process of generating certainty is straightforward. Two conflicting forecasts, each with large variances, are likely to be considered unreliable sources of information if presented to decision makers, but

---

[3] Some estimates of uneconomic investments, even if halved, suggest an industry and regulatory failure of massive proportions. They also remind us of the maxim that the hardest system of regulation to reform is a grossly inefficient one since there are so many interests threatened by improvements in efficiency.

Chapter 14: Stranded Investments Costs

those two forecasts, each with large variances, that are incorporated into a single forecast in a settlement agreement is likely to be accepted as gospel, especially if it has passed through three or four stages of review before it gets to the decision makers. What starts out as conflicting weak guesses evolves during the decision process into key planning parameters.

The mistakes on which I want to focus are (1) the ambiguity of regulatory objectives and (2) the over-estimation of utilities' monopoly powers.

The Problem of Ambiguity in Regulatory Objectives

Although economic regulation of natural monopolies is almost always justified as a method for reducing the inefficiencies of an unregulated monopoly, regulation when actually imposed seldom results in an unqualified pursuit of efficiency. Instead, it tends to embody conflicting or partially conflicting objectives. Furthermore, the debates over the appropriate uses of utility monopoly powers never end. Objectives seem to multiply as the regulatory system ages.

The proponents of regulating in pursuit of economic efficiency argue for elimination of monopoly rents and pricing in pursuit of efficiency, but more powerful voices urge the creation of monopoly rents and use of them to purchase desirable social objectives.[4] The existence of the latter perspective suggest that the harm of an unregulated monopoly is not that it generates monopoly rents and inefficient prices but that it allows those rents to accrue to owners of the regulated monopoly rather than to a "responsible" social agency for disbursement in pursuit of worthy social objectives.

The multiplicity of objectives and the intensity of their pursuit is familiar to every student of regulation as it is practiced in the U.S. They include

---

[4] It is noteworthy that the most publicized response of the U.S. DOE to the California Public Utilities Commission's April 20, 1994, order announcing its intent to require retail wheeling by California utilities was an expression of concern that the Clinton Administration's climate challenge initiatives were threatened by widespread retail wheeling. The environmentalists were also quick to note that many of their valued programs were also threatened, and some California legislators were quick to express concerns about the threat to certain social welfare and tax programs.

not only the cost of explicit programs added to utilities' agendas, but also efficiency reductions caused by mispricing.[5]

One of the more dramatic example of costs imposed on utilities that would not be imposed on unregulated firms operating in competitive markets is the cost of ensuring relative price stability. In a market in which demands fluctuate, the regulator of a strong monopoly has a choice of stabilizing price and satisfying all demands by varying outputs, of stabilizing outputs and satisfying all demands by varying prices, or of selecting some mixture of the two. U.S. regulators have had a tendency to favor price stability and, as a consequence, to support the maintenance of capacity needed to obtain such stability. Gaining price stability for many by maintaining large "reserve" margins of generating capacity is expensive compared to using hedging contracts for those who are willing to pay for such stability. Much of the existing reserve generating capacity will be "excess capacity" when competitive prices prevail. These stranded investments will develop because competitive power markets will allow price changes to ration capacity.

The Problem of Fading Monopoly Powers

The second mistake by regulators and industry executives when evaluating forecasts has been to over-estimate the strength and durability of utilities' monopoly powers. Both regulators and utility managers (and consumer spokespersons, environmentalists and public-sector planners) have failed to appreciate the erosion of the industry's monopoly powers. In particular, there has been a failure to recognize the degree to which the natural monopoly of generation has been propped up and re-enforced by tying arrangements with the natural monopolies of distribution and transmission. The development of robust, extensive, efficient systems of transmission combined with efficient systems of communications within and among utilities eroded the natural monopoly of generation.[6]

---

[5] A recent example of these costs was given by Bernard M. Fox in "The 1994 Electric Executives' Forum," cited above.

In Connecticut, approximately 20 percent of our electric rates are the result of public policy initiatives, including deferred recovery of cost previously incurred, the cost of demand-side managements programs, front-loaded payments to trash-to-energy plants and other private power producers, low-income customers' subsidies, and taxes such as a gross receipts tax. (p.39.)

Chapter 14: Stranded Investments Costs

There is a certain irony in the phenomenon of PUCs urging utilities to pay above-replacement-cost prices for non-utility generating services while counting on the monopoly power of utilities to pass those costs to customers. In reality, they were not counting on monopoly powers over generation; they were counting on natural monopolies of transmission an distribution, supported by state powers, to protect the financial positions of regulated firms. They overestimated all these powers. Accustomed to thinking of the vertically-integrated utility as a natural monopoly, they were blinded to forces that are remarkably clear with hindsight.

Necessary Conditions for Successful Economic Regulation

The following conditions are necessary for successful economic performance under the U.S. system of economic regulation.

◆ The structure of the industry must be consistent with the underlying economics of the industry.

◆ Regulatory practices must be consistent with the underlying theories of economic regulation.

◆ Practices of regulation and the industry must be consistent with widespread public values.

◆ Monopoly power of the regulated firm must be strong enough to survive the pricing errors that are inescapable features of the system.

Today the industry fails all four tests. First, the structure of the industry depends for its justification on a belief that generation is a natural monopoly, which it hasn't been since transmission and communication systems made it possible to coordinate many generators over a large area.

Second, underlying theories of regulation call for limiting economic regulation to natural monopolies.

---

[6] For a discussion of early attempts to exploit the potential of transmission networks to reorder the structure of the industry, see William J. Hausman and John L. Neufield, "Public Policy and the Structure of the Electric Power Industry: Some Less from the Past," *Regulatory Responses to Continuously Changing Industry Structures*, (Institute of Public Utilities, Michigan State University, East Lansing, MI 1993).

Third, widespread public values support the use of impersonal, competitive markets to allocate goods and service when such markets can promise efficient results.

Fourth, the monopoly power of electric utilities derived from transmission and distribution services is not sufficient to protect the regulated firm from making errors in pricing generation services inherent in the fully distributed cost model of pricing.

Utility managers and regulators less intent on using monopoly powers for purposes other than efficiency might have noticed the fading of their monopoly powers much sooner. Similarly, regulators and utility managers less captured by the mindset of detailed and extensive adjudications might have elevated their level of abstraction when viewing the industry and recognized more clearly its inter-relations with its economic environment.

Interim Summary

Utility managers and regulators, confident of their abilities to use utilities' monopoly powers to recover costs created by forecasting failures, accepted risks in the 1970s and 1980s that more observant managers and regulators might have avoided. Contemplating those failures and the proposition that the art of economic forecasting has not improved noticeably in the last two decades should warn us not to place any more faith in long-term forecast of demands and fuel prices on which estimates of stranded investments are based than we ought to have placed in the long-term forecasts of demand and fuel prices made in the 1970s and 1980s.

Evaluation of Forecasts of Stranded Investments

Most of the current estimates of stranded investments are based on simple forecasts that current price-cost relations in the industry will prevail for the remaining life of the stranded investments. Such forecasts cannot be taken seriously since the restructuring itself will make dramatic changes in price-cost relations. Useful estimates of stranded investments must be based on forecasts that recognize that generating assets are durable: They can render services for many years, and during that time the national economy will move from recession to prosperity and back to recession several times.

Stranded Investments is a Number to be Created Not a Number to be Discovered

While it is not useful to design the industry's transition to competitive power markets to minimize stranded investments costs, it is useful to exploit opportunities to improve efficiency and simultaneously reduce stranded investment costs and/or raise funds for stranded investment compensation. There may be many such opportunities. Two working hypotheses are proposed herein.

◆ Any action that improves efficiency and simultaneously reduces the magnitude of stranded investments (or raises funds to be used to compensate stockholders for stranded investments) is an action that ought to be taken.

◆ The size of stranded investments next year is dependent on regulatory policies adopted today.

A useful starting point for analysis is that current pricing policies reduce the efficiency of the industry to below its potential. Some of these policies will not survive the transition to competitive bulk-power markets and others, if not changed, will hold the efficiency of the new, competitive industry below its potential. One critical question is whether regulators are flexible enough to grasp these opportunities, and whether they can cooperate in doing so.

The Analysis Builds upon Four Propositions

Proposition One. If stranded investments exist, there also exists a potential for repricing utility services so that benefits to all consumers can be gained. In fact, it is likely that the larger stranded investments and the "bad contracts" problem, the greater is the opportunity for quick efficiency improvements.

Proposition Two. The task of regulators is not to block the transition until utility stockholders have been compensated for stranded investments. Instead, it is to exploit efficiency improvements quickly and compensate stockholders by sharing the benefits of efficiency improvements with them.

Regulators have at their disposal many ways to reduce the dimensions of the stranded investments problem and to do so rather quickly. Once they accept restructuring as inevitable and a problem to be managed rather than a force to be resisted, they can begin to reduce the magnitude of the problem. Cooperation among state regulators and between federal and state regulators is important if gains are to be large and predictable.

Proposition Three. Competition is not the objective of restructuring; improvements in efficiency are the social objective. Competition is only one instrument; reformed regulation is a second instrument. Competition, unaided and undirected, will not produce as many benefits as competition made efficient by improved regulation.

Proposition Four. Faulty pricing of regulated services has brought into existence more generating capacity than is currently needed for reliable and efficient services, and faulty pricing will entice even more capacity into existence if such pricing is permitted to continue.

Pricing to Improve Efficiency and Reduce Stranded Investments

The pricing problem can be usefully separated into the problem of overpricing and the problem of underpricing. It is convenient to deal with the problem of overpricing first.

The first principle of regulatory pricing is that prices charged by a regulated firm should not encourage a customer, or a subset of customers, to turn to an alternative supplier when that supplier incurs a cost to produce the service greater than the cost at which the regulated firm could produce it and the customer's (or customers') departure requires the customers who remain to bear higher costs.

Regulators, with their concentration on fairness rather than efficiency and who have only a limited knowledge of the economic costs of providing services, find it difficult to avoid this mistake. They appear to have made it by perpetuating the famous, or infamous, interstate settlements procedures in the telecommunications industry. A good case can be made that this process saddled AT&T with cross subsidies and inefficient prices that encouraged "high cost" competitors to come into the industry.[7]

Regulators made a similar mistake when they chose to "encourage" cogeneration and small power production by selecting high-end-of-the-range estimates of avoided costs. In New York the legislature, with its

---

[7] See for example, Peter Temin, *The Fall of the Bell System* (Cambridge University Press, Cambridge, 1987). In qualification, a strong case can be made that AT&T's monopoly over long distance services was doomed the day microwave communications were proven to be reliable for a broad range of services, since that technology narrowed the range of tolerable regulatory error below that needed to make the U.S. system of regulation work.

6¢ law, proved once more the importance of the working rule that "all useful point forecasts will be wrong."

To recapitulate, whatever solution is chosen for the stranded investments problem should not be one that attracts additional investment into the industry and intensifies the problem. For example, it appears that even though existing excess capacity exceeds need, some utilities will find it necessary to add peaking capacity soon. This demand for capacity derives directly from the policy of protecting ratepayers from high prices that on-peak costs would create if prices were set to recover such costs. Since such cost are not recovered in peak periods, they must be recovered at other times. Because peak period prices are kept well below peak-period costs—sometimes they are set to recover only a very small fraction of peak-period costs—there exist only weak incentives to conserve during such periods. The consequence is excess generating, transmission, and distribution capacity in the system. When competitive markets are permitted to set such prices, high prices will develop during peak-demand periods.

Challengers vs. Defenders in the Generation Industry

One big source of stranded investments is capital intensive base-load generators, both nuclear and coal. That problem deserves analysis.

Consider an existing plant (herein called the defender plant and assumed to be utility owned) and a replacement plant (herein called the challenger plant and assumed to be owned by a NUG) where the average cost of the challenger is lower than the average cost of the defender. If the choice were between the two plants, each pricing its services at average cost, and shareholders required to suffer losses involved in closing down the defender, it is obvious that utility consumers would be better off if the defender was closed down and the challenger replaced it.

If, however, the choice were between the two plants, each pricing its services at average cost, and customers of the defender were obliged to ensure that defender's stockholders recover their assets, consumer interests would call for the replacement of the defender with the challenger if present discounted value of the cost savings produced by the challenger over the remaining life of the defender were greater than the cost of compensating the defender. In effect the best strategy for the customers of the defender would be to "buy out" the defender, close the plant and buy from the challenger.

If, on the other hand, the discounted present value of the expected stream of cost savings were not as large as the cost of "buying out" the defender, the best strategy for consumers would be to postpone period by period the elimination of the defender until the discounted cost benefits did exceed the "buy out" cost.

This logic creates a paradox. If the customers "buy" the defender's plant, and thereby become the new defender, and the defender's plant has short run marginal costs (SRMC) of less than the challenger's average total cost (ATC), rational consumers will not close it down; they will operate it at prices at or below the ATC of the challenger and recover some of what they paid to buy it. Only if the ATC of the challenger is below the SRMC of the defender will they close the plant and invite the challenger to enter the business.

Once is it is recognized that the defender can produce for the remainder of the life of his assets at SRMC and that the challenger must expect prices equal to or greater than his ATC to be attracted into the industry, an alternative comparison of consumer benefits becomes possible. first, it is obvious that if the defender (either the original one or the new one who is the agent of customers of the old one) is required to bear the capital loss of his operations, and if he has the ability to do so, he can set a price just at or below the ATC of the challenger and deter entry for the remainder of the defender plant's life. At the end of his plant's life he would surrender the market to the challenger. His customers remain better off with his prices than with the entry of the challenger at the challenger's prices. This approach allows defenders who bears risks of capital losses to minimize such risks. Its by-product is that challengers are deterred from entry until the defender's SRMC rises to equal the challenger's ATC. Society consequently gets the lowest possible real cost of production.

When the case in which the defender's customers have agreed to carry all capital risks is re-analyzed, it is apparent that customers who compensate the defender for the difference between his ATC and the ATC of the challenger do not do so in order to entice the defender to leave the industry and bring the challenger in, but to create circumstances in which the defender can get his prices below the ATC of the challenger and deter, or at least delay, the entrance of the challenger.

Chapter 14: Stranded Investments Costs

This high level of abstraction overlooks many dimensions of the stranded assets investments problems. Two in particular deserve note.

◆ A well-established principle of utility pricing is that demands of customers reflect the "law of demand," that is, customers tend to buy more when the price is low than when it is high. *Ceteris paribus*, there is a two-part tariff that will increase consumers' welfare over that produced by a volumetric rate if the energy component in the two-part tariff is below the volumetric rate and equal to or above SRMC, and if the second part of the tariff does not vary with actual takes of energy. Consumer welfare is maximized when the energy charge is equal to SRMC. One widely understood objective of regulators is to find such two-part tariffs.

◆ A less well-emphasized principle is a corollary of the first one, namely, that customers who can be served under two-part tariffs will always prefer the technology with the lowest marginal cost if the ATCs of the technologies are equal.

Since the current transition is one in which most challengers, in the next few years at least, will use natural gas as their generating fuel and most defenders are using coal or nuclear energy, the defenders will tend to have lower marginal costs over most of the ranges of their outputs. Consequently, defenders with an ATC equal to that of challengers have a competitive advantage.

Recovering Stranded Investments

If stockholders of a defender plant must be compensated for capital losses caused by selling at a price equal to the challenger's ATC when the defender's ATC is higher, where might the needed monies come from? The answer proposed here is that three sources have a potential to generate funds to pay for stranded investments and simultaneously contribute to improved economic efficiency.

Example One. Owners of durable capital operating in competitive markets normally expect to recover their capital and profits in peak-demand periods. In a society with business cycles, that means prolonged periods of little recovery of capital followed by prolonged periods of relatively large recovery.[8] Regulators and industry executives should not try to evaluate stranded investments under the assumption that current excess capacity will last forever and that the slow growth of the past few years will last forever. On the contrary, one objective of regulators and industry executives ought to be to work off excess capacity so that the proportion of time when generating assets can be expected to recover some fixed cost is increased.

Example Two. Many generators are demand constrained rather than price constrained, that is, the owners would prefer to run them more at prevailing prices but can't find a buyer for the power. Once efficient, competitive markets are created, such markets will provide generator owners with unlimited buyers for their generators' services. As long as the system has excess capacity, the market-determined prices are not likely to cover full costs, but the opportunity to run at the chosen capacity factor will tend to lower the unit cost of the low marginal cost plants below what many of them could obtain in the currently Balkanized industry. As excess capacity is worked off, either by economic growth or by the closing of uneconomic plants, market-determined prices for generation output will rise towards and to the ATC of challenger plants.

Example Three. Some utility assets have a market value higher than their book value. This provides an opportunity to improve the efficiency of utility operations by creative accounting. It does not necessarily further efficiency for firms to buy power at market-determined prices that are efficient and simultaneously move that power over transmission and distribution lines that are underpriced by efficiency standards. Regulators can serve the efficiency objective by increasing the prices of transmission and distribution services, especially transmission services, when those prices reflect embedded costs that are less than replacement costs.

First, regulators can price transmission and distribution services, especially transmission services, on a replacement-cost basis and compensate owners on an original-cost basis and use the difference to compensate utility-owners of stranded generation or to pay compensation to NUGs

---

[8] Economic cycles other than business cycles can create opportunities for some fixed-cost recovery. Examples are daily, weekly and annual cycles. The importance of annual cycles for some business is illustrated by specialty stores that recover most of their fixed costs in the Christmas season and garden stores that recover most of their fixed costs in the planting season.

for accepting revised power sales contracts. Second, regulators can cease recording depreciation allowances for all transmission and distribution assets whose book values are below challenger costs. An equivalent increase in depreciation can be taken on generating assets whose book values are above their market values.

Obviously, regulators can pursue both policies. Furthermore, depreciation can be stopped on each generator whose book value is less than its market value so as to concentrate capital recovery on plants with book values above their market values. They can also match generators with book values above their market values and generators with book values below market values into a package that has market and book values equal and insist that stockholders who want to be compensated for the stranded investments of the generator with book value above market value accept the package as a deregulated asset. A careful investigation of the value of the land and other assets included in the package should obviously precede such package creation. A recognition that many plants may have a plant life far longer than their book life can also help in making a market valuation of generating assets.

Such policies might tax the ingenuity of accountants, but with cooperation from federal regulators state regulators might find that such policies could add to efficiency and reduce the scope of more painful forms of "taxation" that may be required to raise funds for stranded investments compensation.[9]

Conclusions

Absent a change of mind by the Congress, the electric industry will be restructured to create a competitive generating market. This change will also demand a restructuring of the jurisdictional responsibilities of regulators. The only sustainable objective for the industry and its regulators is to bring the operation of the industry into conformity with underlying economic realities and facilitate orderly restructuring so the industry can also operate in conformity with current theories of regulation, with current public values and with the capabilities of the regulatory system.

---

[9] Compare "Consistency in approach is needed so that there are no wide differences in the treatment of stranded investment from state to state. Such differences could artificially upset the competitive playing field, especially between adjoining states." Charles F. Goff, "The 1994 Electric Executives' Forum." p.44.

To accomplish these objective and to maintain public support for the restructuring, it is necessary that net benefits flow to all customers from the restructuring during the restructuring, not merely after it.

Given the diversity among states and the need for a high degree of uniformity in neighboring states, industry leadership in crafting plans for programs to master the stranded investments problem must be a high priority.

CHAPTER 15

Preparing for the Inevitable: The Nationalization of the U.S. Nuclear Industry in a Competitive Electricity Market

Dr. Edward P. Kahn
Economist
Lawrence Berkeley Laboratory

The Incompatibility of Competitive Electricity Markets with Nuclear Power

This paper[1] argues that the U.S. electricity industry faces one enormous barrier in its transition to a competitive structure, namely the lack of a coherent policy toward nuclear power. The forces pushing the electricity industry away from regulated monopoly and toward a competitive structure are arguably irreversible. This movement is part of a world-wide phenomenon that is affecting electricity markets to varying degrees in all countries. I will comment on the international experience briefly below. In the U.S., the forces pushing toward a competitive structure are particularly strong. The most potent of these forces is the gap between high tariffs for industrial customers and low-opportunity costs to meet their demand. This point has been argued at length, and it will not be pursued here.[2]

In the U.S., the dialogue about the transition to competition has focused on the unproductive use of euphemisms, particularly the notion of "stranded assets".[3] This notion loosely describes the gap, referred to above, between

---

[1] A similar though less developed version of this argument appeared in E. Kahn, "A Modest Proposal: Nationalize the U.S. Nuclear Industry to Foster Competition," *The Electricity Journal*, v. 7, no. 5 (1994) 44–47.

[2] This point is made routinely by spokesmen for industrial energy consumers; *see*, for example, the paper by Anderson in this volume. Cohen and Kihm (1994) relate this point specifically to the high costs of nuclear power.

[3] *See*, for example, H. Cavanaugh, "Big Issues for '94 – Top Execs Trim to Compete, But Uncertainties Slow Them Down," *Electrical World*, v.208, no. 1 (1994) 5–15.

the procedures used to determine rates and the competitive price. The rate determination process depends upon asset valuations that are excessive relative to the competitive prices. In some areas of the country, there are commercial contracts and regulatory bargains that contribute further to the misalignment. These facts are undeniable, although there are legitimate disputes about the magnitudes involved. Much less attention, however, has focused on the qualitative differences among the assets and commitments that are currently stranded. In this paper, I will argue that nuclear power, alone among the "stranded assets," is a barrier to competitive markets because there is no private sector solution to the problems it poses. This qualitative distinction is so important, and so widely ignored, that it will inevitably call for federal government intervention to resolve.

This argument is structured in the following fashion. In the rest of this section, I lay out the evidence for the qualitative incompatibility of nuclear power with a competitive electricity industry. This evidence includes the experience of other countries with this problem, the inherent uncertainties of nuclear fuel cycle economics, and the safety risks posed by ignoring the problem. In the second section, I review the Cold War origins of nuclear power. This history sheds important light on why the technology evolved as it did, and why federal intervention will be necessary to manage it now that the Cold War is over. The third section addresses these management tasks and their financial implications. Nationalizing nuclear power will be substantially less costly than the savings and loan problem, but will pose much more difficult management tasks. The fourth section summarizes the argument.

Preliminary Evidence: U.K. and Argentina
By now it is well known that the British government was unable to privatize the nuclear industry in the U.K. during the reorganization of the electricity industry.[4] It is worth focusing on the scenario in a little detail to appreciate exactly what was involved and why it failed.[5]

---

[4] J. Hewlett, "Lessons from the Attempted Privatization of Nuclear Power in the United Kingdom," *Energy Sources*, v.16, no.1 (1994) 17–37.

[5] I follow D. Newbery and R. Green, "Regulation, Public Ownership and Privatization of the English Electricity Industry," Conference on International Comparisons of Electricity Regulation, Toulouse, France, 1993, and J. Chesshire, "Why Nuclear Power Failed the Market Test in the UK," *Energy Policy*, v. 20, no. 8 (1992) 744–754.

Chapter 15: Preparing for the Inevitable

The original 1988 proposal for a privatized electricity industry in England and Wales specified a duopoly in which the nuclear assets were to be bundled with 60% of the conventional generators. The remaining capacity was to be sold as a separate company. The proposed company with the nuclear assets, what subsequently became National Power, would have had a near-monopoly position in generation. The British had previously privatized industry under such terms. Presumably it would have made the nuclear assets more financially attractive to be bundled as proposed.

Decommissioning and fuel reprocessing costs were to be covered by a fixed-price contract with British Nuclear Fuels Ltd (BNFL). In light of the obligation that would have been placed on BNFL by this arrangement, it re-estimated its future costs in early 1989. These turned out to be about ten times greater than previous estimates. The financial community expressed considerable doubt about the feasibility of selling shares in the proposed National Power. The managers of the nuclear assets requested government guarantees for the commercial and technical risks associated with the proposed structure. The government refused to provide these, and all the nuclear assets were withdrawn from the proposed private company. They were placed in a government-owned company, called Nuclear Electric, where they remain. The privatization of a reduced National Power and the smaller company, Power Gen, proceeded.

As of 1993, the performance of Nuclear Electric has improved significantly compared to its level before the restructuring. Its market share has increased from 16.5% in 1988/89 to 23.7% in the second quarter of 1993.[6] It is also reported that this improved performance has been accompanied by declining costs, indicating a substantial productivity increase. This record has stimulated discussion of another attempt to pursue privatization, this time with Nuclear Electric as a stand-alone company.[7] Despite this renewed optimism about nuclear performance, however, nothing has changed with respect to private sector risk-bearing.

The record in other countries is not nearly so rich and informative. Relatively complete restructuring of the electricity industry is still

---

[6] Office of Electricity Regulation (OFFER), Pool Price Statement, 1993.

[7] *The Economist*, "No Thanks," April 9, 1994, 57–59.

somewhat rare. Both Norway[8] and Chile[9] have effected such transitions, but in neither case were there any nuclear assets in the generation mix before the restructuring. In the case of Norway, the new competitive structure has not been accompanied by widespread privatization. In Chile, all generation assets are now in private hands. The case with the most parallels to the U.K.'s is Argentina's.[10] Here a thorough reorganization of the electricity industry took place in 1991–92. Almost all the thermal and hydrogeneration assets have been sold to private investors. The nuclear units remain under government ownership. In the Argentinean restructuring, no generating company has a large share of the market.

The limited experience of competitive restructuring in electricity and nuclear power suggests that there is an incompatibility. As competitive pressures increase in other countries, this conclusion will be tested. In both Sweden[11] and Canada[12] competitive pressures are growing in the presence of a very considerable base of nuclear generation. These countries will provide interesting tests of the incompatibility hypothesis starting from the initial position of government ownership. Government ownership of the electricity industry enables much greater flexibility for restructuring.[13] Where private ownership is pervasive, elaborate compensation schemes

---

[8] E. Hope, L. Rud, and B. Singh, "Markets for Electricity: Economic Reform of the Norwegian Electricity Industry," SNF Working Paper No.12/1993, Bergen, Norway, 1993.

9 S. Bernstein, "Competition, Marginal Cost Tariffs, and Spot Pricing in the Chilean Electric Power Sector," *Energy Policy*, v. 16, no. 4 (1988) 369–377.

[10] I. Perez-Arriaga, The Organization and Operation of the Electricity Supply Industry in Argentina, Energy Economic Engineering Ltd., 1994.

[11] L. Hjalmarrson, "From Club Regulation to Market Competition in the Scandinavian Electricity Supply System," in R. Gilbert and E. Kahn, eds., *International Comparisons of Electricity Regulation*. Cambridge University Press (to appear).

[12] Ontario Hydro, "Ontario Hydro and the Electric Power Industry: Challenges and Choices," Exhibit No. 2.1.16 in Ontario Energy Board HR 22, 1994.

[13] Newbery and Green (1993) refer to this as the "option value" of government ownership.

are required to restructure assets. Governments can simply write down losses on uneconomic assets. In a setting of private ownership, compensation for "stranded assets" becomes a principal point of negotiation.

The incompatibility of nuclear power with a purely commercial structure in electricity can be illustrated in a number of ways. In the U.K. the issue seems to have focused primarily on the risks associated with the back end of the fuel cycle. The financial community could not put a price on these risks. It was simply unwilling to accept them. Similar problems associated with safety issues appeared at the initial stages of development in the U.S. These are discussed below under the origins of nuclear power.

Fuel Cycle Cost Uncertainties

The costs associated with the back end of the nuclear fuel cycle and the decommissioning of reactors are fundamentally unknown. There are both technological and political elements to the problem. We don't know exactly what society will tolerate in the way of waste disposition and decommissioning. Therefore we cannot know what it will cost. It is this uncertainty which makes it so difficult to place a purely private sector value on nuclear reactors.

In the U.S. considerable effort and financial resources for dealing with waste storage and plant decommissioning have been set aside. The federal government has accepted responsibility in principle for waste storage. J. Holdren[14] reviews the history of government policy toward radioactive waste storage. After years of uncoordinated activity, the Nuclear Waste Policy Act was passed in 1982 calling for the U.S. Department of Energy (DOE) to open, by 1998, a site for permanent deep geologic disposal of both military and commercial high level waste.[15] This legislation also created a structure of fees to be charged to utilities, based on nuclear electricity production, to finance the facilities. In 1985, the Reagan

---

[14] "Radioactive Waste Management in the United States: Evolving Policy Prospects and Dilemmas," *Annual Review of Energy and Environment*, v.17 (1992) 235–260.

[15] OTA (1989), cited by Holdren (1992), estimates that spent fuel from nuclear reactors dominates the high level waste stream when measured by radioactivity, but that military waste dominates when measured by volume of material.

administration announced that the two sites originally agreed upon would not be necessary, thereby focussing attention exclusively on the proposed site at Yucca Mountain, Nevada. Progress in developing this site has been slow; it is virtually certain that the 1998 date will not be met. The result is increasing tension between the utilities and DOE. Of the $8 billion set aside for nuclear fuel disposal, some might be devoted to alternative approaches.[16] Given the political tensions surrounding the waste disposal issue, it is hard to see what will be feasible and what it will eventually cost.

The costs and management of decommissioning nuclear plants are the responsibility of the utilities. Because there has at least been some experience with decommissioning nuclear plants (primarily the original small-scale experimental facilities), there is less uncertainty than in the case of waste disposal costs. A funding mechanism has been established to finance the expected decommissioning cost for each plant. The main risk is that the estimated costs are subject to "appraisal optimism," a disease that has routinely affected cost estimation in the nuclear industry worldwide. G. Fry[17], for example, has reviewed the existing evidence on decommissioning costs and found no evidence of scale economies. Yet all the cost estimates on which the financing is based rely on scale economies. This sort of game will not pass the market test.

Safety Risks

The most fundamental fact of competition in electricity is the pressure that will be created on all costs. Since the driving force behind competition is high tariffs, the utilities will try their hardest to meet the prices of competitors by reducing their own costs. This is normal and expected behavior. In most cases, it is desirable. The exception to the unambiguous desirability of cost reduction involves safety.

Although the U.S. nuclear industry is subject to substantial safety regulation by the Nuclear Regulatory Commission (NRC), the basic facts of regulation suggest incomplete enforcement. Regulation is fundamentally reactive. The ability of any regulatory structure to control behavior is limited. Agencies always have less information than the companies they

---

[16] E. Bretz, "Special Report: Nuclear Power," *Electrical World*, v.208, no. 7 (1994) 27–41.

[17] "The Cost of Decommissioning U.S. Reactors: Estimates and Experience," *The Energy Journal*, v. 12, Special Issue (1991) 87–104.

regulate. The modern theory of regulation suggests that these realities be dealt with by creating incentives for the socially desirable behavior.[18] This incentive approach is unlikely to have much practicality in the case of nuclear safety. The problem is that there is a conflict of interest. A safety problem arising from competitive pressures on cost might have been avoided by spending more money on operations and maintenance (O&M). But the cure is precisely what the utilities are trying to avoid in the first place. Therefore the situation resembles a zero-sum game, rather than the positive-sum game usually imagined in the incentive literature.

A safety incident at a U.S. nuclear plant is likely to precipitate a major economic crisis for all the electric utilities with such plants. The accident at Three Mile Island in 1979 had a profound effect on the nuclear industry. The response to that accident was a substantial increase in the safety requirements for existing nuclear plants and for those under construction at the time. I do not think another accident would produce the same result, particularly if it were shown that the accident resulted from inadequate operations and maintenance procedures. An accident related to underfunding of O&M, itself induced by competitive forces, would create strong pressure to nationalize the nuclear industry on very unfavorable terms.

A "safety panic" over the role of nuclear power under competitive conditions would probably end up shutting down a large fraction of the oldest and/or poorest performing nuclear units. It is unlikely that a scenario of this kind would result in a particularly economic or equitable outcome. There would not be much attention to important distinctions concerning the costs and benefits of decisions concerning individual plants.

Financial Distress and the Asset Transfer Process

The competitive process produces winners and losers. Losing firms, those which cannot meet their financial obligations, are reorganized. The underlying assets of these firms are still devoted to useful purposes, but their ownership changes. Trying to imagine how this process will work in the electricity industry is difficult. The problem will be similar to that the British government experienced during the privatization. There will not be any willing buyer for the nuclear assets. The costs of the fuel cycle risks described above cannot be estimated reliably enough for someone to put a meaningful commercial price on them.

---

[18] J. Laffont and J. Tirole. *A Theory of Incentives in Procurement and Regulation.* Cambridge, Mass., MIT Press, 1993.

Accelerating competition may produce a "financial meltdown" for utilities owning nuclear plants. The normal asset reallocation process (i.e., bankruptcy) is unlikely to work. In this case, there is only one solution. Give the assets back to where they originally came from, that is, the federal government.

The Origins of Nuclear Power as a Cold War Technology

Part of the argument for nationalizing the nuclear industry rests on a recognition that this industry has, in fact, always been a creature of federal government policy, and therefore must ultimately be accepted as such. The best evidence for this proposition lies in the history of the industry in the 1950s, during the height of the Cold War.

The Soviet Threat

The light-water reactor technology that ultimately became the basis for commercial nuclear power in most countries was one of several competing approaches to nuclear power generation under investigation in the late 1940s and early 1950s. It had been selected by Rickover in 1950 as the basis for the Navy's submarine program. In 1953, the National Security Council decided that it was in the national interest to end experimentation with a variety of technologies and to focus effort on the one approach to reliable nuclear power that seemed technologically feasible. R. Cowan[19] refers to this decision as a classical case of "technological lock-in," meaning that after the commitment of national resources had been made to the light water approach, all other alternatives, which might have been more economic in the long run, were foreclosed.

Why did the National Security Council make this decision in 1953? Perry and his colleagues[20] describe the crucial factors somewhat cryptically as "national prestige and foreign policy considerations." Cowan's characterization is more explicit. The government thought Soviet nuclear power technology would be used to induce the governments of developing countries to align themselves with Soviet economic and political interests.

---

[19] "Nuclear Power Reactors: A Study in Technological Lock-In," *Journal of Economic History*, v.50, no.3 (1990) 541-567.

[20] R. Perry, A. Alexander, W. Allen, P. deLeon, A. Gandara, W. Mooz, E. Rolph, S. Siegel and K. Solomon, "Development and Commercialization of the Light Water Reactor, 1946–1976," RAND Report No. R-2180-NSF, 1977.

The U.S., it was felt, must compete in this dimension with its own nuclear power technology. However unlikely all this sounds from the perspective of the end of the Cold War and the economic disintegration of the Soviet Union, there were reasons to take these arguments seriously at the time.

The Price Anderson Act
The Atomic Energy Act of 1954 contemplated private sector participation in the nuclear power program. The Atomic Energy Commission (AEC) implemented this policy in 1955 through its Power Reactor Demonstration Program (PRDP). The PRDP continued through 1963, testing various technologies other than the light-water reactor and attempting to induce commercial involvement from vendors and utilities. A crucial step in this process was the passage of the Price Anderson Act of 1957. This legislation followed Congressional hearings during which spokesmen for utilities and vendors testified that without legal limits on the liability of private firms in case of an accident, commercial participation in reactor development would cease. The ability of private sector insurance markets to underwrite the risks of a nuclear accident was insufficient to reassure the potential participants. Therefore, the federal government asserted its right to impose a liability limit.[21]

The Price Anderson Act has been the subject of numerous reviews in Congress. Its constitutional status was tested, and affirmed, by the Supreme Court in 1978. As part of this continual re-examination of the liability issue, an implied federal commitment to underwriting potential liability claims became increasingly explicit. In the Supreme Court opinion upholding Price Anderson, Justice Burger, writing for the majority, cites this backstop commitment by quoting congressional opinions uttered in 1965 and 1975.[22]

The Lessons of History
The point of revisiting these early episodes in the history of the U.S. nuclear industry is to clarify why we have an economic policy problem with this technology. The answer is simply that the technology was promoted, at considerable expense, for non-economic reasons. Therefore solutions to the economic problems created now that the original motivations are no longer relevant cannot rely on private market processes exclusively.

---

[21] H. Green, "Nuclear Power: Risk, Liability and Indemnity," 71 *Michigan Law Review*, 1973.

[22] Duke Power Company v. Carolina Environmental Study Group, Inc., 1978.

Financing and Managing the Nationalization

Now let us suppose that a policy consensus existed that the U.S. nuclear industry should be reorganized under explicit federal responsibility. Achieving such a consensus would, of course, be difficult, given the enormous size of the financial and managerial tasks involved. Hopefully, a policy consensus would not be achieved as a result of an accident-induced "safety panic." Regardless of how it came about, however, a reorganization would involve answering three fundamental questions. First, what should the price be for compensating investor-owned utilities for taking over these assets? Second, how should the federal government finance the transfer? Third, how should the industry be managed under federal responsibility? In this section, I will outline the considerations that should go into answering these questions, and suggest some guiding principles.

What Should the Purchase Price Be?

Nationalization of the nuclear industry means purchasing the assets from their current owners, most of which are investor-owned utilities. The actual transaction is likely to be much more complex than a simple purchase because the plants will continue to operate and their output will probably be used by their former owners.[23] Neglecting other financial aspects of the ultimate transaction, it would be necessary at the start of any financial restructuring to set a purchase price for the nuclear assets. The upper bound on such a price would be the depreciated book value of the plant. This is the price, for example, that a regulatory commission might allow a utility to recover when a nuclear plant is shut down in advance of its expected lifetime. In early shut-down cases, the regulator sets a multi-year amortization period for the payment of this price which typically does not include interest payments on the unamortized balance.[24]

---

[23] This would not necessarily be the case if a larger restructuring of the electricity industry also occurred. If the electricity industry were vertically deintegrated, as in the U.K. and the other countries discussed above, then the government would sell the output to a regional pool.

[24] If interest were paid, then the arrangement would not differ substantially from the standard rate-based treatment of the plant except, of course, that it would not be operating.

An argument can be made that the pricing formula for a nationalization of nuclear assets should take into account a variety of issues such as the vintage of the plant and its record of performance. Vintage effects can be captured in a simple formula, such as setting the price equal to some percentage of depreciated book value. Incorporating performance measures would require a more complicated formula. The desirability of accounting for performance is to hold utility investors financially responsible for management of the plant.

It is important to recognize, however, that precision in a pricing formula is impossible. The purchase price will be a negotiated result that is part of the overall settlement. It will be subject to political influence, since nationalization is fundamentally a political process. A few simple guidelines will shape the negotiation. The most fundamental of these guidelines is that nationalization is a sharing of responsibility among the investors, the utility customers and the federal government. It is likely that investors could suffer very substantial losses in a "safety panic" scenario, or in a "cut-throat competition" scenario, since their exposure in many cases is very large.[25] The motivation for utilities to participate in the nationalization negotiations is that their expected outcome would be superior in such cases. This does not mean that they should expect the upper bound price.

A plausible price outcome would involve a roughly equal sharing of cost among the three parties. This would mean investors would recover two thirds of undepreciated book value on average, with the plants performing well doing better and the plants performing less well doing worse. In this price scenario, the government would buy the assets at the negotiated price and sell the output back to customers for half that price.[26] This approach would result in an equal sharing of the burden, on average, among the three parties.

The financial viability of any nationalization scenario depends critically on how the federal costs would be funded. Various alternatives are reviewed in the next section.

---

[25] Moody's Investor Services, "Nuclear Power-A Current Risk Assessment," 1993.

[26] Strictly speaking, the resale price to utility customers would be their share of the fixed costs computed as suggested, plus the costs of operation. This should result in competitively priced electricity, or at least, nearly so.

Federal Financing Options

There are a variety of ways in which the federal share of the nationalization cost could be financed. None of them are likely to be politically popular. The simplest, and probably least acceptable, procedure would be to pay the federal costs out of tax revenues. Under current fiscal constraints this option might only be available in a "safety panic" scenario. Such a scenario would be similar to the atmosphere surrounding the savings and loan financing. It is prudent to consider other scenarios.

The current style of federal budgetary negotiations favors sectoral solutions to sectoral problems, rather than allocating scarce revenues from broadly based sources such as income taxes. Since nuclear power is an electricity problem, the source of financing nationalization under this theory would have to come from the electricity sector in one way or another. The most direct approach would be to impose a "user fee" on electricity sales. This would be analogous to the nuclear levy in the U.K. The order of magnitude of such a tax would be about 2.5% of electricity industry revenues.[27] This is much less than the nuclear levy in the U.K., which is about 11% of the sales price to final consumers.[28] Nonetheless, a tax is a tax, and politically difficult.

Another alternative financing vehicle involves the disposition of hydroelectric power from federal dams. This power, which typically represents about 5% of total electricity sales, is currently sold under highly preferential terms to municipal and cooperative utilities. It represents a substantial subsidy to this sector. It was thought desirable to subsidize alternatives to investor-owned utilities at the time these arrangements were created. Throughout the history of the electricity industry in the U.S., the government-owned segment has been characterized as providing a kind of yardstick competition for the investor-owned segment. Even today, the threat of municipalization acts as a competitive constraint on the pricing policies of franchised investor-owned utilities. As

---

[27] Assume that the net federal cost would be about $50 billion. This amounts to an average depreciated book value of about $1500/kW for about the approximately 100 GW of capacity, split three ways. If $50 billion were financed with long term special agency bonds whose total annual cost (interest and principal) would be $5 billion. Spreading this over annual industry revenues of about $200 billion, results in a 2.5% tax rate.

[28] Newbery and Green, 1993.

competitive forces begin to erode the structure of regulated monopoly, it may be appropriate to rethink some of these arrangements. A fully competitive industry is incompatible with subsidies to certain segments. In countries where the transition to a competitive electricity market has been made, hydroelectric resources have either been privatized (i.e., in Chile and Argentina), or their output is being sold at marginal cost prices (Norway). A federal policy that simultaneously reorganizes the nuclear industry and removes the subsidy related to hydroelectric would eliminate two distortions and create a self-supporting financing vehicle.

While this alternative has economic and financial appeal, the political barriers to its implementation are large. The communities that benefit from the existing subsidy system will resist losing this benefit. There is a regional mismatch between the areas where nuclear assets are concentrated and those where federal hydroelectric resources are located. Therefore financing a nuclear power nationalization from federal hydro revenues would involve considerable transfers between regions.

Management Issues
It is reasonable that there would be concerns about the efficiency of a nationalization of the nuclear industry. Government management of complex technological enterprises has a poor record of performance. In the U.S. nuclear industry, the record of government enterprise is arguably worse than the performance of the private sector. In light of this, why would nationalization represent an improvement?

A responsible management plan for a nationalized nuclear industry must rely strongly on private sector participation. Among the U.S. utilities that own and operate nuclear plants today, there are some that have superior technical capability. These should be induced to participate in the management of a nationalized industry as contractors. Similarly, among the engineering firms that supply services to the nuclear industry today there is substantial expertise, which should also be brought to bear. Other proposals to reorganize the nuclear industry also rely on this notion.[29] Additionally, there is considerable relevant nuclear expertise in other countries. Both

---

[29] G. Rothwell, "New Nuclear Power Plants Require Changes in Utility Structure," Stanford University Center for Economic Policy Research, Publication No. 361, 1993.

France and Japan have extensive nuclear power assets, which bear a strong technological kinship to U.S. technology. Companies from these countries should also be encouraged to participate in the management.

The incentive problems that raise safety concerns today, as competition is increasing, would also be issues under government ownership or private contracting. To encourage efficiency in the contractor management of nuclear assets, financial rewards are desirable. Such reward structures might lead to questionable operating procedures. Without excessive competitive pressure, such arrangements might be more stable than the status quo.

Summary

The argument presented in this paper is incomplete in many respects. Its fundamental points, however, are simple and difficult to deny. First, nuclear power represents a major, if not the major, source of the "stranded asset" problem. Second, the risks associated with nuclear power are difficult, if not impossible, to price in the private marketplace. This means that private sector solutions to managing nuclear power will not be forthcoming. These facts could be suppressed as long as regulated monopoly protected utilities. Competition exposes these facts. The third fundamental point is that competition will increase nuclear power safety risks. This follows from the relentless pressure on costs that competition creates. Safety regulation is only an imperfect protection from this increased risk. Nationalization of the nuclear industry will almost certainly occur in the unfortunate "safety panic" scenario that would follow from an accident. In the more likely "financial meltdown" (or "cutthroat competition") scenario, a nationalization outcome is also likely. Therefore, prudence suggests we start preparing for these contingencies at least intellectually.

Part 6

The Electric Industry in Transition

INTRODUCTION

The Role of Technology in Serving the Customer's Needs

Arnold R. Adler, P.E.
ASME Legislative Fellow
NYS Legislative Commission on Science and Technology

The movement to increased competition in the electric power industry is not just a function of the economic and regulatory climate of the times, but also of the past and continuing role of technology in providing electric power customers and producers with competitive options.

The three papers in this chapter address three different technological aspects of this transition. While the areas of technology under discussion differ widely, they all embody an assumption of economic competitiveness as a driving force for their future viability and application.

The three papers provide insight into the following technological areas:

◆ Integrated resource planning (IRP)

◆ Electric power generation

◆ Customer-based information systems technology

In each of these areas, options of great technical and economic complexity must be considered in the evolution of a successful competitive strategy that fairly meets and satisfies customer expectations and at the same time achieves reasonable business goals.

Important interconnections tie these three papers together despite their diversity of subject matter. Dr. Frank Kreith provides insight into the planning requirements for shaping future energy policy, as well as with an understanding of the renewable energy options. His paper also highlights the opportunities for improved energy efficiency and how to deal with the social costs in the evaluation process. Fundamental to achieving these future policy objectives is the selection of power generation options available to meet electrical needs in an economically sound and qualitatively acceptable manner. Anthony R. Armor's paper provides this necessary overview of power generation technologies currently in hand or likely to be available during the next decade. Without a technical understanding of these options and their economic performance and environmental characteristics, it would be difficult to arrive at any meaningful policy decisions capable of practical implementation. It is also this diversity of generation options that will ultimately be a key factor in fostering competition in the electric industry. It is, therefore, appropriate that the third and final paper in this chapter deals with this competition. Edward M. Smith brings to us his first-hand experience in the United Kingdom with the application of information technology in the development of new products and services based on customer influence and freedom of choice.

With respect to the energy policy issue, Dr. Kreith indicates that IRP or Least Cost Planning (LCP) is mandated by the 1992 Energy Policy Act and is already practiced at some level by the majority of the states. An underlying assumption in least cost planning is that we understand and agree on the nature and costs of the supply-side and demand-side options that must be evaluated in such studies. Social and environmental effects and associated costs can become significant factors in this and call for an in-depth technical understanding of the systems under consideration. Despite this need for technical analysis, we are often confronted with widely varying information about such systems. Much public information appears to come from proponents and opponents with a personal interest in the alternatives under discussion.

Because of the long-term impact of the decisions that are made as the result of integrated resource planning, it is important to support technical efforts to establish a degree of standardization with respect to applicable criteria and methodology, as well as to maintain access to independent sources of impartial expertise for those charged with making public policy decisions. Such assessments need to be able to take account of the risks and uncertainties inherent in the quantification of the performance, cost, schedule, and scaling factors associated with such technical options at different stages of product development and maturity. Dr. Kreith's paper contains quantitative data and information on methodologies that should prove useful in structuring objective evaluations.

There appears to be general agreement that the transition to a competitive electric industry is well under way and that among the major segments of

Introduction: The Role of Technology

the electric industry—generation, transmission, and distribution—it has manifested itself most strongly in the area of generation and will continue in this direction. While there are regulatory and commercial reasons for this, it also appears obvious that there are more technical choices in existence and under development, with respect to the generation option. Armor's paper looks in some detail at generation technologies in the near term (through the year 2005) and also discusses alternative technologies likely to play a greater role in the generation mix thereafter. Armor's paper should be required reading for those interested in understanding where these technologies stand and what they have to offer. There appears to be little question that the gas-fired combustion-turbine/combined-cycle will be the leader in both U.S. and global generation additions in this decade. However, as Armor notes, coal remains the fuel of most existing powerplants in the U.S., and "clean coal technology" developments, including advances in scrubber performance and reliability, make pulverized coal a viable choice for many generation companies. Also, the successful operation of demonstration projects in atmospheric and pressurized fluidized bed powerplants has led to the planning of relatively large (>250 MW) commercial plants, making this a technology much in evidence in the near-term generation market. Significant technical developments are also taking place in distributed generation, including fuel cells, and in renewables such as photovoltaics, windpower, and biomass. Extensive penetration of distributed generation and renewables into the generation mix is unlikely to occur until after 2005.

The third and final paper in this chapter is based on Smith's experience in a competitive electricity market, the privatized United Kingdom electric industry. It is a thought-provoking dissertation on how the convergence of information/communications technology with the technology of electricity supply is creating a highly competitive new market for customer-responsive products and services. Smith points out the opportunity that information technology provides to establish a dynamic interface between customer preferences and electric network operations. He makes interesting comparisons between the evolution of competition in the telecommunications industry and what is happening today in the electricity industry. Smith cautions us against seeing this convergence of new technologies as creating incremental changes in an electricity market and industry that we know and have become comfortable with over the years. He provides a convincing argument that the technological innovation taking place, and the information-based knowledge available to us, requires businesses to recognize and be prepared to participate in a new and different competitive market if they want to survive.

Let me conclude that I learned a great deal from the papers in this section, and I found that there was more here than could be learned from a cursory perusal. The three authors did more than serve up technical facts; they provided much food for thought. It may also be appropriate to note that past predictions about the role that specific technologies will play in the future energy market have not always been precisely on target. Those of us who lived through the nuclear optimism of the 1950s and the anticipated demise of combustion-turbine/combined-cycle powerplants due to the Fuel Use Act restrictions of the 1970s have lost some of our hubris when it comes to forecasting the future. But even when things do not turn out the way they were forecast to be, it is not necessarily due to technology letting us down, but rather because we have gained additional insight. We also recognize that technology, while important, is just one factor in a complex equation determining how the electric industry will serve the future power needs of the American public.

CHAPTER 16

The Role of IRP and Conservation in Electric Utility Transition

Frank Kreith, Doc.Sc., P.E.
American Society of Mechanical Engineers
Legislative Fellow, National Conference for State Legislatures

INTRODUCTION

The previous papers have discussed the future of the electric industry from economic, legal and political points of view. These aspects of energy use are important because on the one hand using energy fosters material well being, while on the other the public has concerns about the finiteness of oil and gas resources, disposal of radioactive wastes from nuclear power plants, acid rain, the greenhouse effect and pollution in general. I am an engineer and my perspective is shaped by thermodynamics and engineering analysis. I see a technical option that would simultaneously minimize both the supply and pollution problems, namely, to improve our energy use efficiency. Unfortunately, this country's recent record in improving its energy use efficiency is not very good.[1] In the decade following the Arab oil embargo, U.S. industry reduced the energy required per unit output of GNP by 20%. At the same time Japanese industry achieved a reduction in unit energy consumption of 50%. The Japanese have continued to improve their energy use efficiency, while in the United States during the past few years, the energy use per unit of GNP has begun to rise again. I want to discuss some measures to reverse this trend.

Energy Requirements
In order to assess the future of the electric industry, decision makers need to have good information. The late Kenneth Boulding, a good friend of mine, often said that one of the problems with decisions is that all of our knowledge is of the past and all of our decisions are about the future. Previous energy-use projections have been quite poor and decisions about the future cannot be any better than the information on which we base decisions. In 1973, the late president Richard Nixon proclaimed Project Energy Independence, saying "By 1980, we will be self sufficient and will not need

The Electric Industry in Transition

to rely on any foreign energy." When I participated in President Carter's policy review of energy in 1976, the president expected that by the year 1995 20% of our energy would come from renewable sources. President Ronald Reagan was quoted in the *Washington Post* of February, 1980, as claiming that "Alaska has a greater oil reserve than Saudi Arabia."

The track records of scientists and economists are not much better than those of politicians. In 1972 the Ford Foundation supported a project designed to estimate the future energy needs of this country.[2] The prestigious group assembled for this study used a supply-side approach. Figure 16-1 displays the projection of the Ford Foundation and the actual energy use in this country, shown by the dots below the zero energy growth scenario projected in the study.[3] There is an enormous difference between the projected and actual energy use.

I am not sure what all of the reasons are for the lack of realism in the projections made by politicians, as well as by scientists, about the future needs of energy. It is clear that unless we improve our predictions about energy needs, our decisions on how to meet these needs will be poor. The Carnegie Foundation has recently conducted a two-year study of

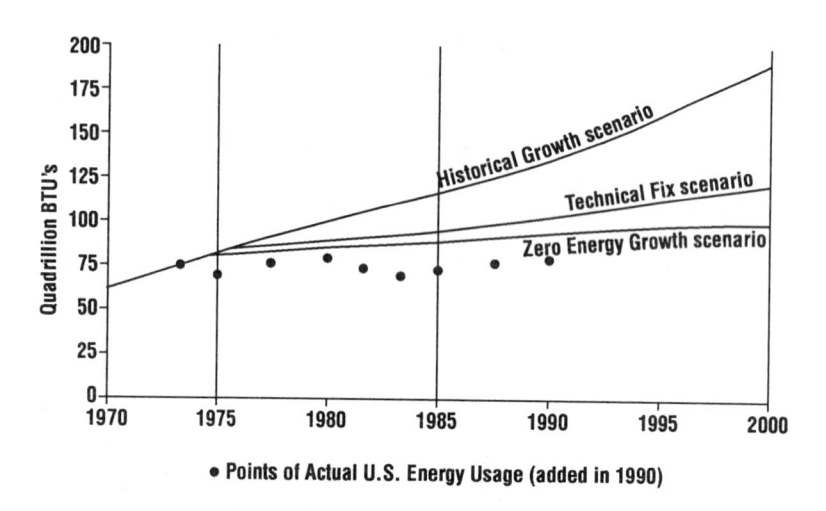

• Points of Actual U.S. Energy Usage (added in 1990)

Figure 16-1. Comparison of Predicted Energy Scenarios* and Actual Energy Use.

Source: Energy Policy Project.
*Ford Foundation Energy Policy Study, A Time to Choose, published 1972.

science and technology and the state governments.[4] Some of the foremost authorities on science and engineering, as well as politicians, participated in this study, including Dr. Erich Bloch, the former Director of the National Science Foundation; Chancellor Donna Shalala of the University of Wisconsin; the former governor of Ohio, R.F. Celeste; Governor Lawton Chiles of Florida; and many others. One of the foremost recommendations of this study was that we must establish a technical information service for state governments, which will be called upon to make some of the most important decisions about energy facing this country in the future.

My discussion of the future of the electric industry in the next few years will be predicated on certain premises and values that are widely accepted today. First, energy use is directly related to environmental quality. The production, distribution and consumption of energy has a direct and measurable impact upon public health and environment. Second, energy efficiency is key to our national competitiveness and to a sustainable future. This recognition requires that integrated resource planning be used in selecting technologies to meet our future energy needs. Our energy resources are finite and the more efficient we become as individuals and as a society, the fewer resources will be required to sustain and improve the quality of life. The higher the energy use efficiency, the easier will be the transition to new generation technologies, whatever they may be, and to the deployment of new energy resources such as natural gas and renewables. Third, government has a role in the operation of energy markets. World energy markets are not free markets and depend on geopolitical and social values. In the U.S., it is widely recognized that job creation is intimately linked to the cost and type of energy used and that government regulations and tax policies have a profound influence on the future direction of energy development and environmental protection.

The previous papers have discussed the impact of deregulation, retail wheeling and stranded assets. The next paper will describe new energy production technologies, such as natural gas cogeneration cycles. I would like to emphasize three points in my paper.

◆ The roles of conservation and renewables

◆ Integrated resource planning

◆ The social costs of energy

Conservation and Renewable Energy

Figure 16-2 shows the per capita energy consumption for some states in 1988. As one can see, on a per capita basis New York uses less energy than any other state in the country.[5] This is in no small measure due to the mass transit system in New York City. However, there are enormous opportunities for both improving energy efficiency and reducing per capita energy consumption all over the country. Figure 16-3 shows opportunities for improving energy efficiency in the residential appliances sector.[6] The figure shows the energy requirements of average models on the markets for refrigerators, air conditioners, water heaters and ranges, the energy requirements of the best commercial model on the market and the potential requirements of newer technologies. Figure 16-4 shows the opportunities for energy savings by replacing existing devices such as motors, lighting and water heating by more efficient technologies. The Electric Power Research Institute claims that by replacing existing devices with more efficient ones, the U.S. could cut electricity consumption by over 31%.[7]

It is easy to be dazzled by the potential energy savings offered by technology, but we should realize that this potential is fraught with a great number of uncertainties. What will it cost? How will it change lifestyles? What are the unknowns, such as geopolitical changes that could affect the adoption of particular technologies? Who will provide the initial investments to replace existing devices? With limited funds available, how will energy savings in the future compete with other public goals in the near term?

I believe that structural changes that result in improved energy efficiency will develop in the future. A driving force will be the continued development and diffusion of information technologies. These information technologies will place a premium on exploiting flexibility and the ability to monitor and control production to exact specifications, characteristics that are inherently energy conserving. Twenty years ago, when I joined the Solar Energy Research Institute as chief of Thermal Research, expectations for the deployment and use of renewable energy technologies were exceedingly high. The expectations have only been partially realized. I believe that major stumbling blocks to the implementation of renewable technology were an overly optimistic view regarding their potential and an unrealistic perspective of their costs. Another obvious stumbling block to the implementation of renewables is our country's failure to include the social costs of energy in the price of energy that consumers pay.

Chapter 16: The Role of IRP and Conservation

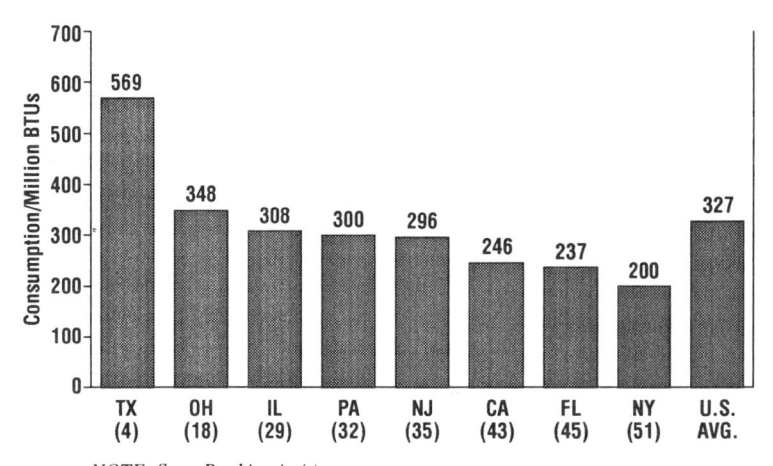

[Bar chart showing Consumption/Million BTUs by state:
TX (4): 569
OH (18): 348
IL (29): 308
PA (32): 300
NJ (35): 296
CA (43): 246
FL (45): 237
NY (51): 200
U.S. AVG.: 327]

NOTE: State Ranking in ( ).
The eight states selected are those with the largest gross state products.

Figure 16-2. Per Capita Energy Consumption for Select States, 1988.

Source: New York State Energy Office, "Energy Policy in the Aftermath of Kuwait," Empire State Report, April 1991, p. 21.

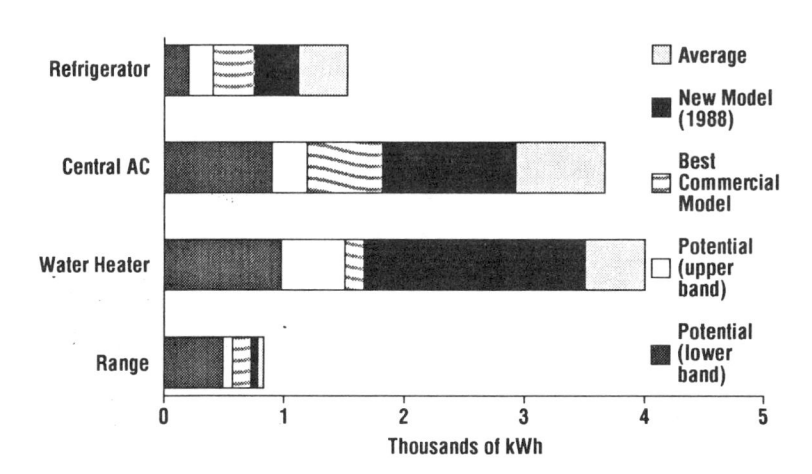

[Horizontal bar chart of Thousands of kWh for Refrigerator, Central AC, Water Heater, and Range, with legend: Average, New Model (1988), Best Commercial Model, Potential (upper band), Potential (lower band)]

Figure 16-3. U.S. Residential Appliances Energy Efficiency Potential.

Peter D. Blair, OTA, 1990

The Electric Industry in Transition

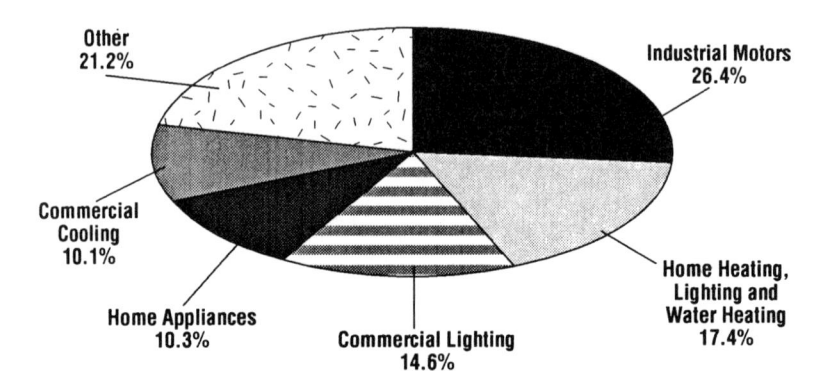

Pie chart:
- Industrial Motors 26.4%
- Other 21.2%
- Home Heating, Lighting and Water Heating 17.4%
- Commercial Lighting 14.6%
- Home Appliances 10.3%
- Commercial Cooling 10.1%

Figure 16-4a. Energy Saving Technologies. Replacing existing devices with more efficient ones could cut U.S. electricity consumption over time by 31.3%, according to a utility group. Here are the group's estimates of which areas the savings would come from.

Source: Electric Power Research Institute

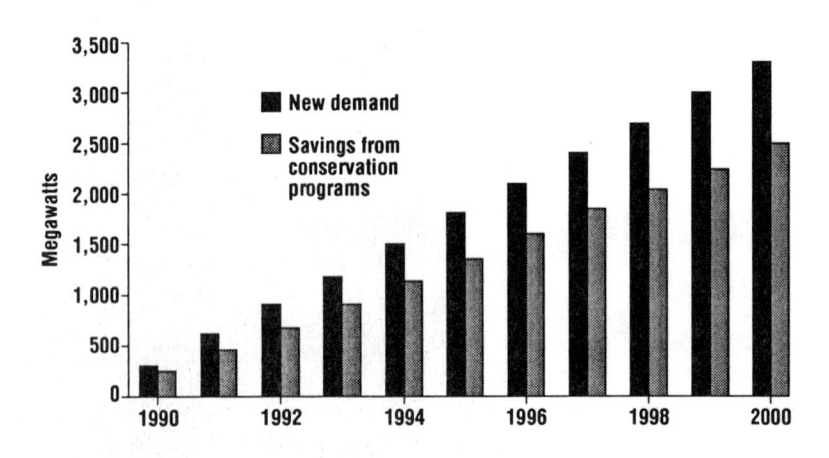

Bar chart (Megawatts, 1990–2000): New demand vs. Savings from conservation programs.

Figure 16-4b. Meeting Demand With Efficiency. Pacific Gas & Electric, the nation's largest utility, expects conservation programs to meet three-quarters of its new demand for electricity by 2000. Here is how its electricity savings will grow along with rising demand to offset the need for power plants.

Source: Pacific Gas & Electric

Chapter 16: The Role of IRP and Conservation

When an energy system operates with solar radiation as its source it is not necessarily renewable. I define a renewable energy technology as one that will return more energy of equal quality during its lifetime than is necessary to construct and operate the system. Under this definition, only four renewable energy technologies qualify at present: hydroelectric power plants, solar thermal heat for domestic hot water or industrial processes; wind energy for electric power production; and biomass, mostly from municipal solid waste, for electric power or heat. On the other hand, photovoltaic technology, although favored by electric utilities, is only barely able to break even at its current state of development in good solar climates. An analysis I have made of the photovoltaic plant in Austin, Texas,[8] showed that it will take almost 20 years before the energy invested in its construction is paid back. If the same power plant were constructed in New York, where the solar climate is less favorable, it would not pay back its energy investment during its lifetime. Hence it cannot be considered a renewable source although it requires no fossil energy input once it is constructed.

Hydropower is an old technology that accounts for an appreciable portion of our energy supply. Solar hot water heaters on the roofs of homes in Israel supply approximately 3% of the total energy need of the country and are economically viable. This technology could be used today, as evidenced by a solar hot water heater in my own home in Colorado that has operated trouble-free for the last 20 years and repaid its energy investment more than four times. Waste-to-energy production from municipal solid waste is practiced in many parts of the country.[9] It is a viable technology that costs a good deal less than recycling per ton of waste disposed. I have discussed its potential and environmental impact in other publications.[10] Here I will restrict my comments to the potential of wind energy.

The lifecycle costs of wind power in favorable parts of the country have dropped dramatically and are today competitive with those of coal-fired and nuclear power plants. Figure 16-5 shows the U.S. wind resource potential.[11] It is apparent that the most favorable sites for wind power are in the central part of the United States. Some of the best sources are in North Dakota and Texas. In the period between 1983 and 1990, lifecycle costs of wind energy decreased from 25 to 30 cents/Kwh to between 7 and 9 cents/Kwh and are expected to drop to about 5 cents/Kwh by the end of this decade. The factors that drove these economic improvements are discussed in Reference 11. From the political perspective, wind energy is not only attractive because of its inherent economics, but also because of its

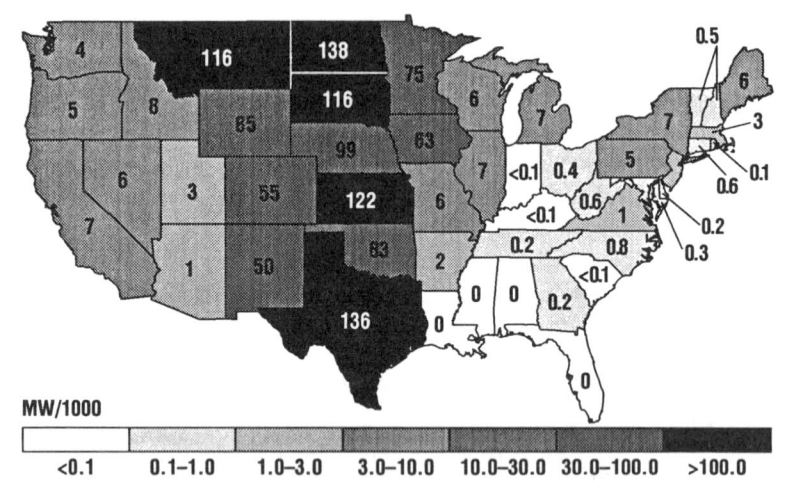

The Electric Industry in Transition

Figure 16-5. U.S. Wind Resource Potential. A 1991 report issued by the Department of Energy's Pacific Northwest laboratory concludes that 0.6% of the contiguous United States contains sufficient wind resources to generate 20% of U.S. electricity demand using today's wind turbine technology. More advanced machines, such as U.S. WINDPOWER'S 33M-VS variable speed wind turbine, could expand U.S. wind electric potential by as much as ten times.

Source: EH1-1/U.S. Windpower, 1991. The numbers indicate the wind electric potential (in thousands of MW_{avg}) that each state's wind resource could provide using advanced technology.

potential for job generation. This is shown in Figure 16-6. According to a study by the World Watch Institute, direct employment from wind energy per Kwh per year is on the order of 542 people, compared to 100 for nuclear or 116 for coal-fired power plants. In an era when generation of jobs is important to all state governments, the enhanced job-generation potential of wind energy makes this technology all the more attractive.

Integrated Resource Planning

Irrespective of whether or not retail wheeling will be implemented, in the long haul integrated resource planning (IRP) will be an important factor in the electric industry transition.[12] After long and arduous debate, the National Energy Policy Act of 1992 was signed by ex-President

Chapter 16: The Role of IRP and Conservation

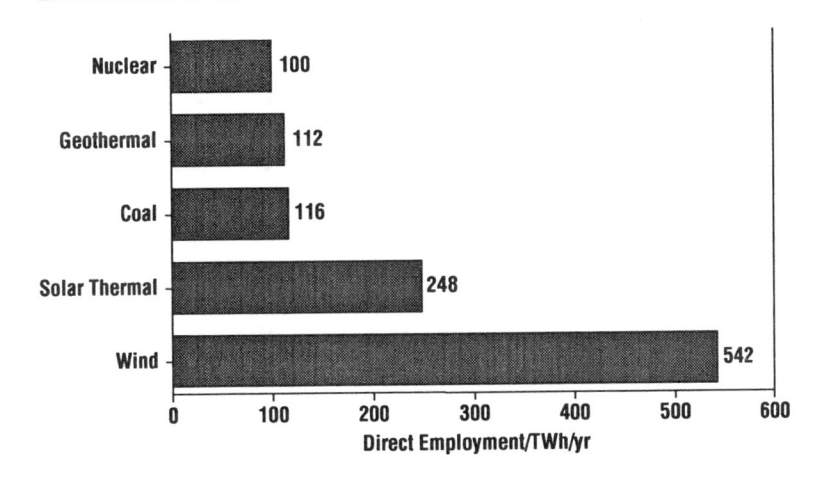

Source	Direct Employment/TWh/yr
Nuclear	100
Geothermal	112
Coal	116
Solar Thermal	248
Wind	542

Figure 16-6. Jobs from Electricity Generation.

Source: Worldwatch Institute, 1991 Paul Gipe & Associates.

Bush on October 24, 1992. It attempts to improve energy efficiency and includes a requirement that all utilities "shall employ integrated resource planning". IRP is defined as "a planning and selection process for new energy resources that evaluates the full range of alternatives, including new generating capacity, power purchases, energy conservation and efficiency, cogeneration and district heating and cooling applications, and renewable energy resources to provide adequate and reliable service...at the lowest system cost" and "...shall treat demand and supply resources on a consistent and integrated basis". The act also requires that the Western Area Power Administration include social costs in the lifecycle system costs.

IRP is the process of simultaneously examining all energy savings and energy-producing options to optimize the mixture of resources and minimize the total costs while including consideration of environmental and health concerns. There is no unique method for IRP, but the following sequence of steps is generally used:[12]

◆ Develop a load forecast.

◆ Inventory existing resources.

The Electric Industry in Transition

- Identify future electricity needs not being met by existing resources.
- Identify potential resource options, including DSM programs.
- Screen all options to identify those that are feasible and economic.
- Identify and quantify environmental and social costs of these options.
- Perform an uncertainty analysis.
- Select a preferred mix of resources, including conservation measures and load shaping, which are treated as synonymous to supply options.
- Implement least-cost mix of supply and conservation options.

Figure 16-7 shows a schematic diagram for an IRP process that includes consideration of demand-side management (DSM) and social costs. It is designed to develop a systematic procedure to evaluate demand-side options, compare these to supply-side options and develop an energy policy that can integrate environmental and social costs. The goal is to develop a long-term energy strategy that will acquire the most inexpensive resources first and internalize social costs of energy (SCE) in the rate structure. It has been suggested that IRP be called least-cost planning (LCP) when social costs are incorporated into the process. However, both IRP and LCP use the same methodology, and the terms are used interchangeably according to EPRI.[7]

The cost (or value) of saved energy (CSE) by installation of a conservation measure such as a high efficiency motor must also be expressed as c/Kwh. It is common practice to use the levelized cost over the lifetime of the system. As shown by Kreider and Kreith,[13] the cost of energy from a conservation system is approximately:

$$CSE = \frac{\text{Initial Cost of Conservation Device} \times \text{CRF}}{\text{Energy Saved Per Year}}$$

CRF, in the above equation, is the capital recovery factor that accounts for the time value of money invested initially. Its numerical value depends on the lifetime of the conservation device, t, and the discount rate, r, or:

$$CRF(r,t) = \frac{r}{1-(1+r)^{-t}}$$

Chapter 16: The Role of IRP and Conservation

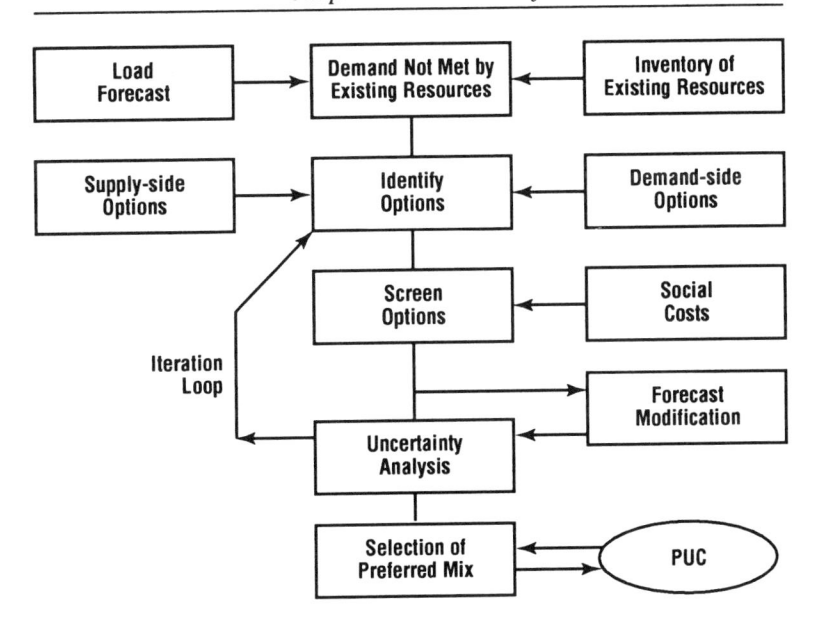

Figure 16-7. Schematic diagram for IRP process.

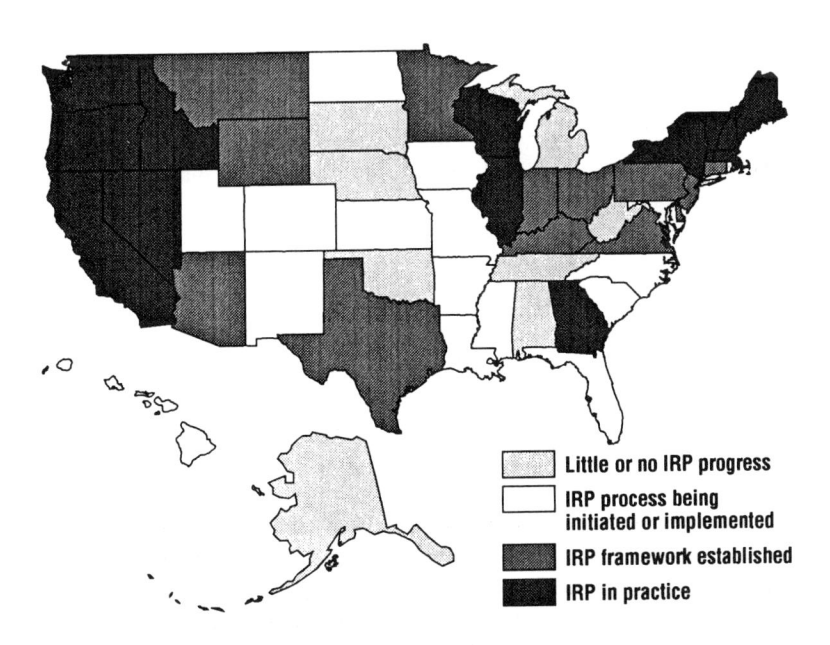

Figure 16-8. IRP Status in the United States.

Table 16-1. Estimated Payback and Energy Cost for Conservation Technologies*

Technology	Life (yr)	Payback (yr)	Cost of Saved Energy (c/Kwh)
Building Insulation	20	1–4	1.5–1.9
Storm Windows	20	5–10	3.5–4.0
Solar Films	3–15	~5	3.2–13
Weather Stripping	2.5	1.6	~5.2
Heat Pumps	15	3–4	2–4
Evaporative Cooling	5–20	—	1.1–3.3
Efficient Motors	~7	1.3**	~5
Heater Insulation	10	1.1	0.9
Low-flow Shower Head	10	0.4	0.4
High-efficiency Refrigerators	20	1.3**	0.7
High-efficiency Fluorescent Lighting	20,000 (hr)	~1**	0.7

* Abstracted from Kreith and Burmeister[14]

** Payback is based on incremental cost.

Table 16-1 shows the energy cost and payback time for some typical DSM measures calculated from data supplied by the Western Area Power Administration (1991). The cost of saved energy (CSE) in the last column was calculated from Equation 2 using a real discount of 3%, a long-term average of the difference between the interest rate and the inflation in the U.S.

According to a study by the Electric Power Research Institute,[7] the DSM measures will reduce electricity demand by approximately 25,000 MW over the next decade. This is equivalent to about 30% of the forecasted new capacity requirements for the same period. EPRI also estimates that DSM activity nationwide will result in a reduction of electricity consumption of about 107 billion Kwh in the year 2000, compared to the amount that would have been used without DSM.

DSM can also delay the upgrading of transmission and distribution systems, an investment that has been estimated to exceed, within the next decade, the investment in new power generation in the U.S. On the state level, DSM is expected to help improve regional economies by reducing the cost of energy. Although there are still concerns about the cost of implementation of DSM, the question for this decade is not whether or not

Chapter 16: The Role of IRP and Conservation

DSM will be used but to what extent it will be used to reduce the need for new power plants and transition and distribution upgrades and expansions.

Ranges of the costs of some existing solar options for use in IRP programs are given in Table 16-2.[12]

Social Cost of Energy

Social costs are difficult to estimate and vary from place to place. One of the pioneers was the New York Public Service Commission, which estimated in 1989 the economic cost of mitigating the residual air emission from a "base" coal power plant that barely meets federal New Source Performance Standards (NSPS) and used that figure as the social or externality cost as shown in Table 16-3. An overview of how other states in the U.S. incorporate externalities in the IRP process is presented in

Table 16-2. Estimated Cost Ranges of Some Solar Options for IRP

System	Cost of Energy in c/Kwh
Advanced Windows (H&S)	1.1–5.0
Daylighting and Controls (H&S)	2.0–4.0
Solar Domestic Hot Water (H&S)	4.0–16.0
Solar Process Heat (Larson)	3.5–5.2
Photovoltaic DC (1990) (H&S)	25.0–35.0
Solar Thermal AC (Kreith, Larson)	8.0–15.0
Wind AC (Kreith, Larson)	4.7–7.2

Table 16-3. New York Externality Cost Estimates[12]

Externality	Emission from NSPS Coal Plant (lbs/Mwh)	X	Control Cost ($/lb)	=	Mitigation Cost (c/Kwh)
Air Emission					
SO_2	6.0		0.416		0.25
NO_2	6.0		0.92		0.55
CO_2	1820.0		0.00055		0.10
Particulates	0.3		0.167		0.005
Water Impacts	NA		NA		0.100
Land Use	NA		NA		0.400
Total					1.405

Reference 12. It was found that the externality estimates for the New York bidding process are lower than estimates presented by Koomey.[15]

Koomey recently surveyed available studies on the external costs of electric power from fossil power plants in the U.S. Excluding CO_2 costs, the externality costs of existing coal-fired power plants were found to range from 1.9 to 3.5 c/Kwh (excluding results for California that were three times as high and an early EPRI study that gave only about half of the above values[12, 15]). For new coal-fired power plants that meet current emission standards (NSPS), the externality costs excluding CO_2 ranged from 0.8 to 1.5 c/Kwh if the values from California and EPRI are omitted. For an average cost of electric power of 6.6 c/Kwh, externality costs are about 18% for new plants and 42% for older plants without state-of-the-art pollution control equipment. The results of the survey for coal power plants are fairly close to previous estimates by Hohmeyer for externalities in Germany. Excluding California, Table 16-4 gives Koomey's averaged values of the externality costs of gas, oil, and coal-fired electric power plants, including estimates of environmental costs due to global warming from CO_2 emission. To obtain the total energy costs, the social costs must be added to the market price of the energy. The estimates of the California Energy Commission (1990) for the cost of utility-owned electric power from various fuel sources are shown in Table 16-5. It is apparent that the externality costs for fossil fuels are substantial, but externality costs per Kwh of heat from direct use of natural gas are considerably less than from electrical heating. No data on the social costs of nuclear power in the U.S. could be found in the peer-reviewed open literature, and there exists a need to determine the social costs of nuclear power for inclusion in least-cost energy planning.

Chapter 16: The Role of IRP and Conservation

Table 16-4. Summary of Average Externality Costs of Energy for the U.S.

Technology	Average Delivered Cost (1989 cents/Kwh)	Average Externality Cost (1989 cents/Kwh)	Externality Cost as % of Delivered Cost
Existing Steam Plants:			
Natural Gas	6.6	0.78	12%
Oil	6.6	1.67	25%
Coal	6.6	2.94	45%
New NSPS Plants:			
Coal Steam (base load)	8.3	1.51	18%
CT Gas (peak load)	5.5	0.95	17%

Table 16-5. Estimated* Cost of New Electrical Generation in 1987 Constant Dollars[12]

	Low	High
	(in cents/Kwh)	
Nuclear (Pressurized LWR)	5.3	9.3
Natural Gas (Intermediate Load)	5.3	7.5
Hydro	5.2	18.9**
Wind	4.7	7.2
Coal Boiler	4.5	7.0
Natural Gas (Base Load)	4.4	5.0
Geothermal (Flash Steam)	4.3	6.8
Biomass Combustion	4.2	7.9

These estimates do not include social costs.

**High cost is for mini-hydro systems.*

I recently learned of a DOE-supported study of the social costs of nuclear power conducted by the PACE Institute. This study estimates that the social costs of nuclear power in the U.S. were 2.5 c/Kwhr. However, key assumptions made in the course of this study were challenged by DOE experts and the PACE report had not officially been released by DOE at the time this paper was returned to the printer (see Quad Report, vol. 1, No. 4, p. 5, 1993 for additional information). The controversy caused by the PACE study increases the urgency for an objective and peer-reviewed determination of the externality costs of nuclear power that is acceptable to the professional community for integration in the IRP process.

Conclusion

Energy efficiency and economic competitiveness will play important roles in the future of the electric industry. Integrated resource planning (IRP) has been mandated in the 1992 National Energy Policy Act and has been initiated by many states, as shown in Figure 16-8. It should play an important role in future planning, although one must keep it in mind that tax incentives, retail wheeling and environmental regulation will have impacts on the competitiveness and cost of all supply as well as demand-side measures. Also, the imposition of CO_2 or Btu taxes and the way in which the social costs of energy are taken into account could drastically influence the decision process. Use of natural gas and cogeneration systems will increase. Renewable energy technologies will be competitive if social costs are properly evaluated in competition with new fossil supply-side sources, particularly natural gas cogeneration turbines. Good information about future energy needs and flexibility in the decision process will improve the inevitable transition. The U.S. will be forced to make a transition to an energy infrastructure using its own resources and employing technologies that improve energy efficiency and reduce adverse environmental impacts. The question is not whether or not there will be a transition, but how the transition will proceed. The transition will either be gradual and orderly, or it may be forced upon the country after a major crisis. New technologies will help in the transition, but they will not be able to produce a sudden solution or panacea to energy dependency and wasteful energy usage.

Acknowledgements

The author would like to express his appreciation to Mr. Matthew Brown, Mr. Eric Sikkemma and Ms. Jackie Cooper of NCSL and to Dr. Arnie Adler of ASME and Dr. Michael Rice of the NYS Legislative Commission on Science and Technology of the New York Energy Office for reviewing this chapter and providing helpful comments. I have attempted to incorporate their suggestions as much as possible, but the final content of this chapter is entirely the responsibility of the author. Financial support for the preparation for this chapter was provided by the Government Affairs Program of the American Society of Mechanical Engineers, but the views expressed are purely those of the author and are not an official position of the American Society of Mechanical Engineers.

References

1. *Science*, vol. 240, *4856*, p. 1121, May 27, 1988.

2. "Time to Choose", Ford Foundation Energy Policy Study, 1972.

3. J.B. Chaddock, D.S., private communication, 1990.

4. "Science, Technology, and the States in America's Third Century", Carnegie Commission on Science, Technology, and Government, September, 1992.

5. "Energy Policy in the Aftermath of Kuwait", *Empire State Report*, New York State Energy Office, p. 21, April, 1991.

6. Blair, Peter T., OTA, 1990.

7. Lamarre, L., "Shaping Demand-Side Management," *Solar Today*, pp. 11–14, Mar/Apr, 1992.

8. Kreith, F. and Norton, P., "The Potential of Renewable Energy to Reduce Carbon Dioxide Emission," *Energy and the Environment into the 1990s*, ed. A.A.M. Sayigh, Pergamon Press, Oxford, UK, pp. 2804–2811, 1990.

9. Kreith, F. ed., *Handbook of Solid Waste Management*, McGraw-Hill, New York, 1994.

10. Kreith, F., "Solid Waste Management in the U.S. and 1989–91 State Legislation," *Energy, the International Journal*, vol. 17, pp. 427–476, 1992. See also, Kreith, F., "Technology and Policy of Waste Management," *Proceedings of the 85th Air and Waste Management Association*, 92–20.07, Kansas City, MO., 1992.

11. "Wind Energy for the States; A Technical and Legislative Perspective", NCSL/ASME, 1993.

12. Kreith, F., "Integrated Resource Planning", *J. of Energy Resources Technology*, vol. 115, pp. 80–85, June, 1993.

13. Kreider, J.F., and Kreith, F., *Solar Heating and Cooling – Active and Passive Design*, McGraw-Hill Book Co., New York, 1982.

14. Kreith, F., and Burmeister, G., *Energy Management and Conservation*, NCSL, Denver, CO, 1993.

15. Koomey, J., "Comparative Analysis of Monetary Estimates of External Costs Associated with Combustion of Fossil Fuels", Report No. 28313, Lawrence Berkeley Lab., Berkeley, CA, 1990.

CHAPTER 17

Generation Technologies through the Year 2005

Anthony F. Armor
Director, Fossil Power Plants
Electric Power Research Institute

INTRODUCTION

This paper reviews the status and the likely application of competing generation technologies, particularly those with near-term potential. Capacity additions in the U.S. will, in the next 10 years, be based on gas, coal, and to some extent on renewables. Repowering of older plants will likely be increasingly attractive.

Gas turbine-based plants will dominate in the immediate future. The most advanced combustion turbines achieve more than 40% efficiency in simple cycle mode and greater than 50% Lower Heating Value (LHV) efficiency in combined cycle mode. In addition, Combustion Turbine/Combined Cycle (CT/CC) plants offer siting flexibility, phased construction, and capital costs of between $400/Kw and $800/Kw. These advantages, coupled with adequate natural gas supplies and the assurance of coal gasification backup, make this technology a prime choice for green field and repowered plants.

There are also good reasons why the pulverized coal (PC) plant should still be a primary choice for many generation companies. Scrubbers have proved to be more reliable and effective than indicated in early plants. Up to 99% SO_2 removal efficiency is possible. By the year 2,000, 60GW of U.S. coal-fired generation will likely be equipped with flue gas-desulfurization (FGD) systems. Also, the PC plant has the capability of much improved heat rate (about 8500 Btu/Kwh) even with full flue gas desulfurization.

Atmospheric and pressurized fluidized-bed combustion (FBC) of coal offers reductions in both SO_2 and NO_x and also permits the efficient combustion of low-rank fuels. In the U.S., there are now over 150 units generating power and ten vendors of FBC boilers, four of which offer units of up to 250 MW in size.

Gasification power plants exist at the 100 MW and 160 MW levels and are planned for up to 450 MW. Much of the impetus is now coming from the DOE clean coal program, in which three gasification projects are in progress and four more are planned.

In small unit sizes, often suitable for distributed generation, technical progress will be made (although large-scale applications will still remain modest) in renewables (solar, wind, biomass) and in fuel cells. Use of fuel cells promises high efficiencies, low emissions, and compact plants.

Capital costs will remain a determining issue in the application of all these possible generation options.

Background

The U.S. electric utility industry consists of a network of small and large utilities, both private and public. There are more than 3,000, which generate more than 700 GW of electric power. This is by far the greatest concentration of electric power production in the world, exceeding in fact those of the next five countries combined.

The National Energy Strategy of 1991 set general policy for the future. One specific directive was to enhance efficiency of generation, transmission, and use of electricity. There are two key drivers for this directive: one to reduce emissions of undesired air, water and ground pollutants and the other to conserve our fossil fuels. The recent energy tax proposed by the new administration is another result of the desire to move to a more energy-conscious mode of operation. It will clearly give impetus to the use of renewables as significant (although probably not major) future power sources and will encourage efficiency advances in both fossil and nuclear plants.

Management of Existing Fossil Plant Assets

Utilities are now looking at power plants in a more profit-focused manner, treating them as company assets to be invested in a way that maximizes the company bottom line or the profit for the utility. As the average age of fossil units inches upwards, utility executives often ask, "If I invest a dollar into this plant to improve heat rate, availability or some other plant performance measure, will this produce more in base profit to my company than investing in some other plant, or building new capacity,

Chapter 17: Generation Technologies

or buying power from outside?" One important aspect of this new business strategy concerns the "use" that is being made of any particular plant, since increased plant usage implies more company value in that plant.

Here are some measures of plant utilization. Heat rate is of course the quantifiable measure of how efficiently we can convert fuel into MW. It is inherently limited by cycle and equipment design and by how we operate the plant. In a simple condensing cycle the heat rate of a fossil plant cannot fall much below 8500 Btu/Kwh, even if the plant uses supercritical cycles and double reheat of the feedwater.

Capacity factor (CF) is another measure of use that tells us how the plant is loaded over the year. Few fossil units achieve 90% capacity factor these days, and this has an impact on the measured heat rate of the unit. Under ideal conditions for effective asset management (and apart from downtime for maintenance) CF should be close to 100% for the purposes of getting the most out of the plant asset. Market conditions and the reserve margins of the utility often dictate otherwise.

The cost of electricity is a determining factor in how units are dispatched. Electricity cost is largely dictated by fuel cost, which typically makes up 60 to 80% of the cost of operating a fossil power plant. It is interesting that none of the top ten U.S. units in heat rate is among the top ten in electricity cost.

Finally, I have suggested a term called "energy efficiency" that describes how well a plant utilizes the basic feedstock (coal, oil or gas). If we can produce other products from a fossil plant besides electricity, the value of that asset goes up and of course the "effective" heat rate drops significantly.

We are seeing many, perhaps most, of the major U.S. utilities use these measures when taking a close look at their plant assets and judging their bottom-line value to the company. An upgrade or maintenance investment in a power plant will now be determined largely on the return on investment the utility can expect at the corporate level. In order to achieve corporate goals, utility staff will need to have a good handle on equipment life and the probability of failure. They also need options for improving heat rate, for increasing output (by repowering), for improving plant productivity to make the assets competitive.

Clean Coal Technology Development

At an increasing rate in the last few years, innovations that are aimed at reducing emissions through improved combustion and environmental control in the near term and by fundamental changes in the way coal is pre-processed before conversion of its chemical energy to electricity in the long term have been developed and tested. Such technologies are referred to as "lean coal technologies", and constitute a family of pre-combustion, combustion/conversion, and post-combustion technologies. They are designed to provide the coal user with added technical capabilities and flexibility and at lower net cost than current environmental control options. They can be categorized as:

- Precombustion, in which sulfur and other impurities are removed from the fuel before it is burned

- Combustion, in which techniques to prevent pollutant emissions are applied in the boiler while the coal burns

- Postcombustion, in which the flue gas released from the boiler is treated to reduce its content of pollutants

- Conversion, in which coal, rather than being burned, is changed into a gas or liquid that can be cleaned and used as a fuel

Coal Cleaning. Cleaning of coal to remove sulfur and ash is well established in the U.S., where there are more than 400 operating plants—mostly at the mine. Coal cleaning primarily removes pyritic sulfur (reduction of up to 70% of the SO_2 is possible), and in the process increases the heating value of the coal, typically by about 10% but occasionally by 30% or higher. Additionally, if slagging is a limiting item increased MW may be possible, as at one station that increased generation from 730 MW to 779 MW. The removal of organic sulfur, chemically part of the coal matrix, is more difficult but may be possible through using microorganisms or chemical methods; research is underway. Finally, heavy metal trace elements can be removed. Conventional cleaning typically removes 30 to 80% of the arsenic, mercury, lead, nickel, and antimony.

Pulverized-coal fired plants. Built in 1959, Eddystone 1 at PECO Energy was and still is a supercritical power plant with the highest steam conditions in the world. The main steam pressure of this double reheat

Chapter 17: Generation Technologies

machine was 5000 psi when it was built, and its steam temperature was 1200°F. PECO Energy will continue to operate Eddystone 1 to the year 2010. That is an impressive achievement for a prototype unit.

However, the most efficient pulverized coal-fired plant of the future is likely to be a combined-cycle plant, perhaps with a topping steam turbine, as shown in Figure 17-1. With a 1500°F air turbine and 1300°F topping steam turbine, the heat rate of this cycle is about 7200 Btu/Kwh, which is very competitive with any other proposed advanced cycles in the near term.

There has been a perception that the pulverized-coal power plant has come to the end of the road—that advanced coal technologies will quite soon make the PC plant with a scrubber, whose efficiency hovers around 35%, obsolete. This perception is premature. There are good reasons, as we have seen above, why the PC plant should still be the primary choice of many utilities. Up to 99% SO_2 removal efficiency is possible. By the year 2,000, 60GW of the U.S.'s coal-fired generation will likely be equipped with FGD systems.

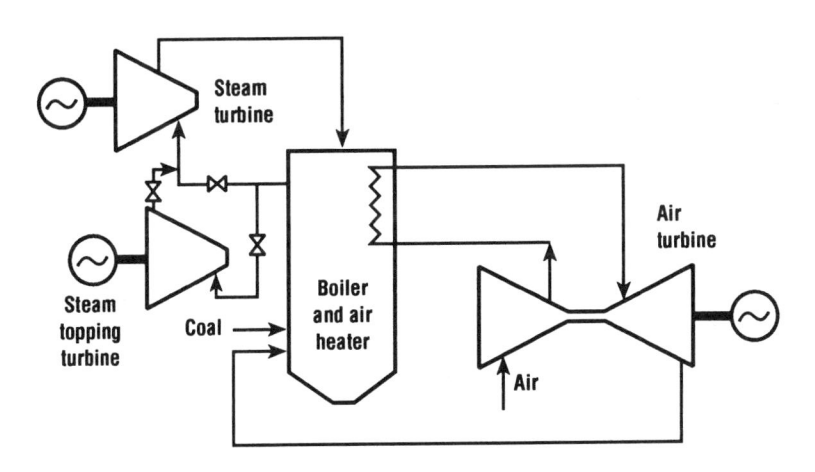

Figure 17-1. A pulverized coal combined-cycle plant with a topping steam turbine has a projected heat rate of 7200 Btu/kWh. The air turbine uses 1800°F air with supplemental firing. The topping turbine uses steam at 1300°F.

Fluidized-bed Plants

Atmospheric fluidized bed combustion (AFBC) (Figure 17-2) offers reductions in both SO_2 and NO_x and also permits the efficient combustion of low-rank fuels. In the U.S. there are now over 150 operating units generating power and ten vendors of FBC boilers, four of which offer units of up to 250 MW in size. The rapid growth in the number of FBC units and their capacity is shown in Figure 17-3.

The demonstration projects at TVA (Shawnee, 160 MW) and Northern States Power (Black Dog, 133 MW) are examples of successful bubbling-bed units. The Shawnee unit now routinely operates at 85 to 90% capacity, having overcome initial fuel-feed problems. The Black Dog unit has been dispatched in a daily cycling mode and has successfully fired a blend of coal and petroleum coke. The focus of AFBC in the U.S. is now on circulating fluid beds (CFBs). The CFB project at Nucla (Colorado Ute) has been successful in demonstrating the technology at the 110 MW level, and planned commercial plants have now reached 250 MW in size. Most fluidized-bed units are being installed by independent power producers in the 50–100 MW size range, for whom the inherent SO_2 and NO_x advantages over the PC plant have encouraged installation even in such traditional non-coal arenas as California.

Pressurized Fluid-Bed Boilers

The future of fluidized-bed utility boilers is evident in the move by several countries (Sweden, Japan, Spain, Germany, U.S.) toward the pressurized fluidized-bed combined cycle design. The 80 MW units at Vaertan (Sweden) and Escatron (Spain) and the 70 MW unit at Tidd (AEP) are already of commercial size, and larger units up to 350 MW are under development. The modular aspect of the pressurized fluidized-bed combustion (PFBC) unit is a particularly attractive feature in that it leads to short construction cycles and low-cost power. This was particularly evident in the construction of the Tidd plant (Figure 17-4), which first generated power from this combined cycle on November 29, 1990. This $185 million project was partially funded ($60 million) by DOE. The heat rate and capital cost of the PFBC plant are forecast to reach very competitive levels, which when combined with shortened construction schedules will put the technology in position to play a role in future generation plans, particularly when modular additions have advantages.

Chapter 17: Generation Technologies

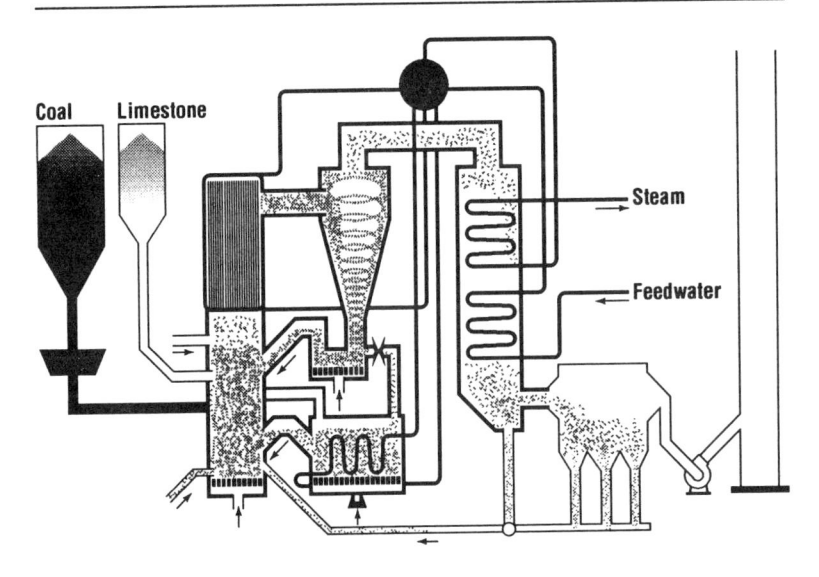

Figure 17-2. The addition of limestone or dolomite to the combustion chamber allows the coal limestone mixture to be burned in a suspended bed, fluidized by an underbed air supply. The sulpher in the coal reacts with the calcium to produce a solid waste of calcium sulfate. The combustion temperature is low (1500F), reducing NO_x emissions.

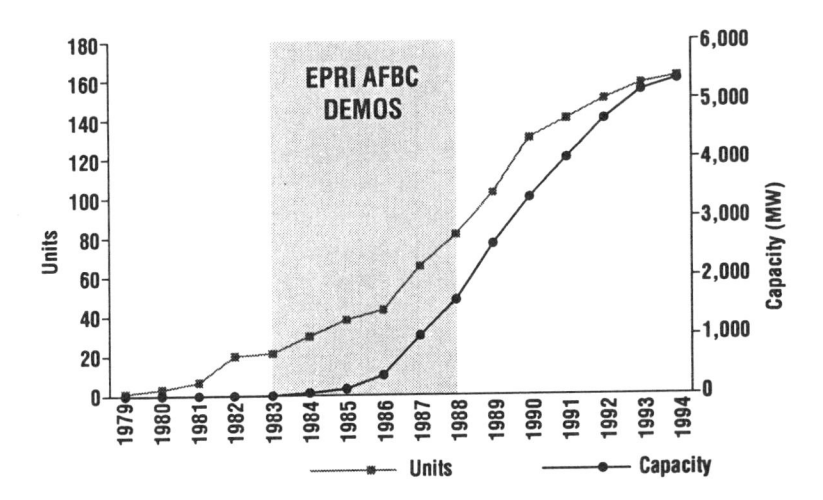

Figure 17-3. The growth in the number and size of AFBC plants has been significant since the demonstration in the U.S. of utility plants at 100MW or more in size.

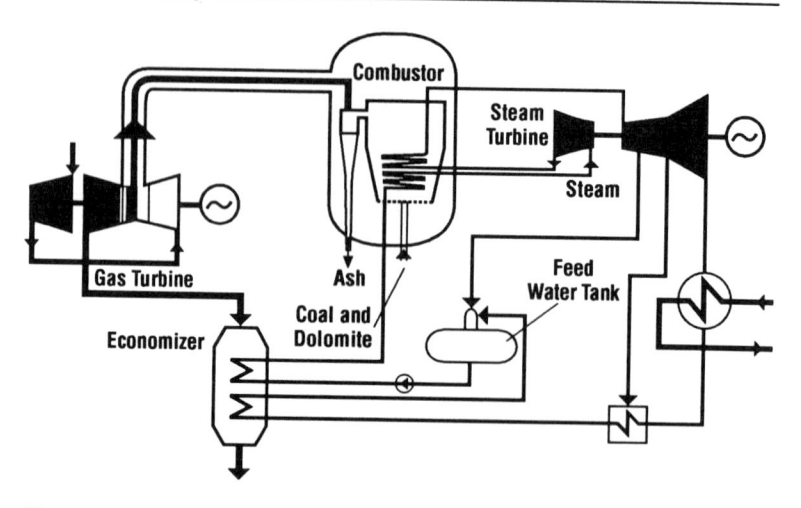

Figure 17-4. Pressurized, fluidized-bed combustor with combined cycle. This 80MW system is now in operation at the Tidd plant of American Electric Power.

Combustion Turbine Plants

Combustion turbine-based plants are the fastest growing technology in power generation. Between now and 2002, natural gas-fired combustion turbines and combined cycles burning gas will provide 50% to 70% of the 900–1000 GW of new generation that will be ordered worldwide. One manufacturer forecasts that there will be 925 GW of new orders worldwide by 2001 and 575 GW installed. General Electric forecasts that combustion turbines and combined cycles will account for 45% of the new orders globally and 66% of the new U.S. orders. Almost all of these combustion-turbine (CT) and combustion-cycle (CC) plants will be gas-fired, creating a major expansion in the use of gas for electricity generation. The Utility Data Institute shows that U.S. utility and non-utility generators are planning to add 97 GW by 2000, of which gas accounts for 37 GW.

EPRI estimates that utilities will install 37 GW and that non-utility generators will install 11 GW of new gas-fired capacity in the United States between 1990 and 2000 (Figure 17-5). Almost all of the utility gas-fired capacity will be combustion turbine-based. While the figures for non-utility generation have not been broken out, it also will be

Table 17-5. Combustion-turbine and combined-cycle plants dominate gas-fired capacity additions. Driven by new high efficiencies, larger sizes, and low capital cost, CTs and CCs predominate for gas-fired capacity.

Gas-Fired Capacity Additions (MW)

Type of Capacity	Installed By 1990	Installed By 2000	Increase 1990–2000
By Electric Utilities			
Steam Generator	100,085	103,187	3,102
Combined Cycle	6,111	17,651	11,540
Combined Turbine	19,898	42,816	22,918
Total	126,094	163,654	37,560
By Non-Utility Generators			
All types of capacity	6,972	18,128	11,156

EPRI Report TR-101239, 9/92

almost exclusively by CT and CC. Worldwide there are even more striking examples. In the United Kingdom, 100% of all new generation ordered or under construction uses gas-fired combined-cycle technology.

Aircraft Technology. In the 1960s, gas turbines derived from military jet engines formed a major source of utility peaking generation capacity. Fan-jets, which replaced straight turbo-jets, were much more difficult and expensive to convert to utility use, and the resulting aeroderivative turbines have been little used. The main reason for the high cost was the need to replace the fan and add a separate power turbine.

As bypass ratios and hence fan power have increased, the most recent airline fan-jets can be converted to utility service without adding separate power turbines. Furthermore, modifications of these engines, with intercooling and possibly reheating, appear to be useful in advanced power cycles such as chemical recuperation and the humidified air turbine. EPRI is managing a major project with a utility consortium to convert advanced aircraft fan-jet engines to high pressure-ratio, natural gas-fired utility gas turbines. Commercial readiness by about the year 1999, very high efficiency, very low NO_x emissions, low water usage, and low capital cost are specific project goals.

Humidified Air Power Plants. EPRI has great interest in what are termed "humidified air" power plants. In these combustion turbine cycles the compressor exit air is highly humidified prior to combustion. This reduces the mass of dry air needed and the energy required to compress it.

The continuous plant cycle for this concept is termed the Humid Air Turbine (HAT). This cycle, using extensive modification of the TPM FT4000, has been calculated to have a heat rate on natural gas about 5% better than the latest high technology combined cycle. The HAT cycle is highly adaptable to coal gasification—especially with quench-cooled gasifiers. Analysis of the integrated gasification HAT indicates that it would have about 10–15% lower capital cost than that of an integrated gasification CC with approximately the same heat rate.

Gasification Plants

One option of particular importance is that of coal gasification. Since the EPRI Coolwater demonstration at the 100 MW level in 1984, the technology has moved ahead in the U.S. largely through demonstrations under the CCT program. Overseas, the 250 MW Buggenham plant in Holland is scheduled to begin operations in 1994, and the PSI/Destec 265 MW and TFCO 260 MW clean-coal demos should be on line by 1996. Thereafter there is a 300 MW plant scheduled for operation in Endesa, Spain, and a 330 MW unit for RWE in Germany (Figure 17-6).

Efficiency improvements in combustion turbines are being made concurrently with the advances in gasification. The new F-type CTs operate at 2300°F, and there will likely be 2500°F machines by the turn of the century. This makes the IGCC a very competitive option.

Distributed Generation

A new approach to meeting cost, environmental, and customer-service objectives is distributed generation, which creates an integrated delivery network employing small, modular generators as well as central stations. Sites and permits of such modular generators (typically of from hundreds of kW to tens of MW) can normally be obtained easily. The plants can be assembled quickly and even relocated should needs change. Strategic placement in the subtransmission and distribution system provides frequency control, voltage regulation, and local spinning reserve capacity; it may allow deferral or avoidance of T&D capacity upgrades.

Chapter 17: Generation Technologies

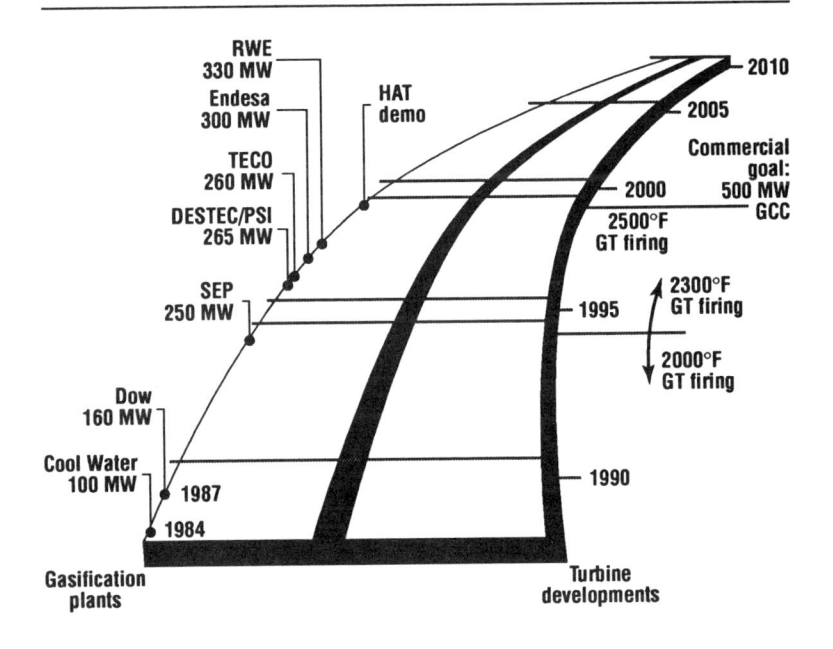

Figure 17-6. Gasification Power Plant Timeline

Some distributed generating technologies—fuel cells in particular—operate quietly with extremely low emissions and are thus well suited for environmentally sensitive areas. Moreover, placement of distributed generators near customer sites enables utilities to offer more reliable service to facilities with critical power needs. Such tailored energy services may prove paramount in meeting competitive challenges from other power producers.

Commercially viable technologies using distributed generation include internal combustion engines and small CTs. Emerging technologies include fuel cells, markedly improved CTs, and improved reciprocating engines.

Fuel Cells

After more than a decade of fuel cell research and development at EPRI, two U.S. manufacturers may be offering commercial molten carbonate fuel cells (MCFC) units within the next three to five years. A utility "buyers group" has placed advance orders for nearly 100 MW of the initial commercial units in exchange for future royalties. As demand grows,

The Electric Industry in Transition

the economies of standardization and mass production will reduce unit costs substantially and expand cost-effective application (Figure 17-7).

If production cost targets are met, MCFC technology could already be cost-effective in some instances. Working with EPRI, the Los Angeles Department of Water & Power, Central and South West Services, and Oglethorpe Power evaluated the benefits of prototype 2-MW fuel cells in various distributed applications. Projected savings ranged from 2 to 71 mills/kWh compared to the costs of power generated at central stations.

EPRI is providing technical management for a consortium of utilities and manufacturers that will demonstrate a commercial prototype of a modular 2-MW molten carbonate fuel cell. This quiet, efficient unit will serve customers of the City of Santa Clara, California. Ground-breaking ceremonies were held in April 1994; commercial generation will begin in 1996.

Renewables

At this time, most attention is focused on wind, photovoltaics, and biomass. Environmentally relatively benign, these technologies will find a small but growing market in specific regions.

Wind. Wind-powered technology is now providing utility customers with electricity and its use is growing despite the expiration of favorable tax credits in the mid-1980s. In California and elsewhere, some 17,000

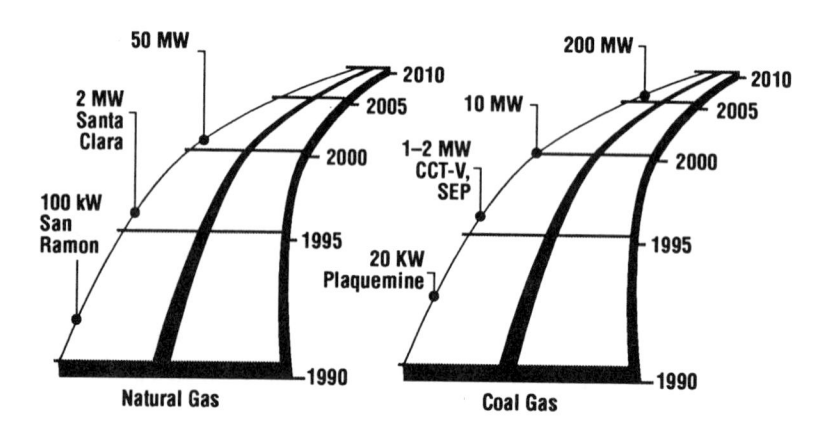

Figure 17-7. The timeline for molten carbonate fuel cells shows that 10 to 50 MW units are possible by the turn of the century.

Chapter 17: Generation Technologies

midsize wind turbines are producing approximately 1500 MW of electricity. The energy costs 7–9¢/kWh, which is competitive with the costs of moderate conventional energy. The total, installed costs for state-of-the-art wind systems range for $1000 to $1200/kW based on name plate rating. EPRI considers wind turbines a viable near-term energy source and is helping to develop an advanced variable-speed wind turbine that promises to lower electricity costs yet further. This machine, the 33M-VS, is designed to generate 400kW of power at roughly 5¢/Kwh cost to utilities.

By incorporating recent engineering developments in advanced electronics, aerodynamics and wind turbine design, the 33M-VS captures significantly more energy than existing constant-speed wind turbines at far lower cost. An advanced power electronic converter allows the rotor and generators to accelerate in higher winds while maintaining a constant frequency output. This feature increases the range of wind speeds (from 9 to 65 mph) over which the turbine operates while significantly reducing the loads to which the turbine is exposed.

Under a joint DOE/EPRI Program, prototype machines will be tested at utility sites, including Green Mountain Power and Central and Southwest.

Photovoltaics. Photovoltaic (PV) cells are made of thin layers of silicon or other semiconductor materials that convert sunlight directly into electricity. Individual cells are small, with typical outputs of 0.6 V and 1–2 A; they are wired in series and in parallel to provide greater voltage and current. A PV plant might contain millions of cells. There are two basic types of PV modules. Conventional modules, called flat-plate modules, collect sunlight unassisted. Concentrator modules use lenses or other devices to concentrate sunlight onto small, highly efficient cells; these modules closely track the sun on special moving assemblies to keep solar rays sharply focused.

There are many reasons why PV technology may have the greatest potential of all the emerging solar technologies for large-scale power generation. PV is highly modular, so systems can be deployed in a wide range of sizes and scaled up as needed, thus reducing technical and financial risks. PV systems accomplish sunlight-to-electricity and dc-to-ac conversion entirely with solid-state electronic components and with few or no moving parts, offering high equipment availability and low operation and maintenance costs. Furthermore, PV systems release no pollutants and, except for land use, they ensure minimal environmental impact.

The Electric Industry in Transition

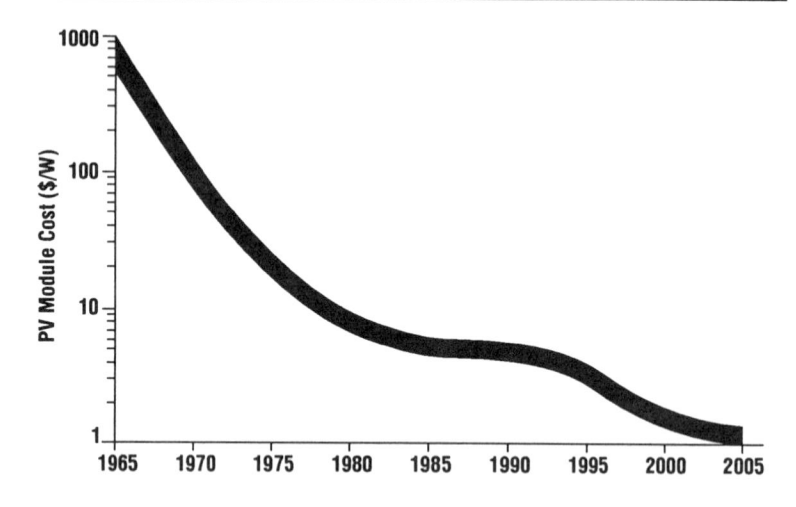

Figure 17-8. Photovoltaic generation costs are predicted to fall to competitive levels by the early 2000s.

PV costs have fallen by a factor of 10 over the past two decades. They now hover in the range of $8000–$12000/Kw for systems of several Kw or more, with energy costs of 30–50¢/Kwh. Costs must drop to the $1500–$2000/Kw and 6–8¢/Kwh range if PV is to contribute significantly to the nation's electrical needs. Future cost reductions are expected to come from module efficiency increases, system design improvements, improved manufacturing technology, and production volume increases (Figure 17-8).

Biomass. Biomass, produced in an economically and environmentally sustainable manner, could realistically be used to supply 50,000 MW of electric capacity by the year 2010 and probably twice that amount by the year 2030.

The U.S. Department of Energy has estimated that at present there are about 1000 biomass-fueled power plants in this country, mainly with capacities of less than 25 MW. About one-third of the power generated is sold to local utility companies, while the remainder serves as self-generation for the industrial facilities creating the wastes. In most instances, the acquisition cost of the biomass fuel has been small; thus

Chapter 17: Generation Technologies

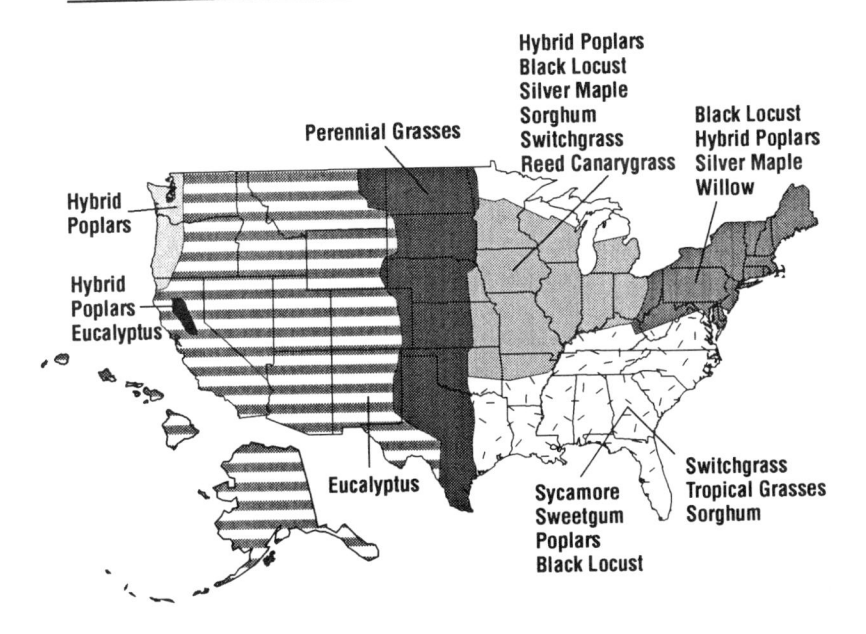

Figure 17-9. Biomass as a fuel for power generation will focus on fast-growing trees and grasses in the future. Today biomass is focused on wastes from agriculture and industry.

the low efficiencies associated with combusting an energy resource with a high moisture content (about 50%) and a low heat content (about 8,500 Btu/dry lb) have not proved a deterrent.

With a diversity of plant species being selected for high yields as well as for drought, blight and pest resistance, the feedstock of biomass power generation is no longer limited to the waste products of the paper and wood industries and agriculture. The biomass feedstock of the 21st century will include hybrid cottonwood, poplar, silver maple, red alder, black locust, sweetgum, eucalyptus, and sycamore trees; willows, reed canary grass, switchgrass, other perennial grasses; and some annuals, such as sorghum (Figure 17-9).

Conclusions

The generation picture for the United States can be divided into two phases: one including the next 10 years and the other, those after 2005. In the near term gas-fired new additions, led by more efficient and reliable combustion turbines, will dominate. Coal-fired plants will be added but in smaller unit sizes than those that characterized the last 20 years, with 400 MW being a likely upper limit. Coal will remain the fuel of choice for most existing generators and provide the majority of generated power.

After 2005 distributed generation, perhaps in the form of fuel cells, will increasingly appear at strategic load points in the electric system. In addition the gas turbine plants of the 1990s will be adapted for coal gasification as natural gas becomes scarcer and more expensive, and the 30 to 40 year old fossil plants that have been the backbone of the utility industry will be retired as more efficient new generation begins to dominate.

Bibliography

1. "GE Forecasts Worldwide Plant Orders of 925,000 MW Between now and 2001," *Independent Power Report*. pg. 1, July 3, 1992.

2. "Plan 97,000 MW of new capacity by 2000: UDI finds," *Cogeneration*, pg. 5, Oct.–Nov. 1991.

3. Armor, A.F., Touchton, G.L., Cohn, A., "Powering the Future: Advanced Combustion Turbines and EPRI's Program," EPRI Coal Gasification Conference, San Francisco, Oct. 1992.

4. Parkinson, J., Suardini, P., "Coal Cleaning to Reduce Sulfur Dioxide Emissions," EPRI CS-5713, May 1988.

5. Couch, G., "Advanced Coal Cleaning Technology," IEACR/44, London, IFA Coal Research, December 1991.

6. Patterson, W.C., "Coal Use Technology in a Changing Environment," *Financial Times Management Report*, FT Business Information Ltd., 1990.

7. Electricity from Sun and Wind, EPRI publication GS 3008.7.91, 1991.

8. A "How to" Primer for Biomass Resource Development, EPRI TR-103439 Dec., 1993.

CHAPTER 18

Business and Technology Change towards the Year 2005: Redefining the Electricity Industry's Customer Markets, Products, and Services

Edward M. Smith
Chief Executive Officer/Director
Perot Systems Europe (Energy Services) Ltd.

SUMMARY

Products and services falling within the electricity industry's traditional franchised (licensed) area of operations are being exposed to significant competitive pressures arising in part from global trends towards industry privatization and re-regulation. The immediate effects of these changes can be seen in customers being afforded a freedom of choice in selection of their preferred electricity supplier and in the increasing level of recognition amongst suppliers over the measure of customer influence in determining the shape of new products and services.

Convergent advances between information and technology, telecommunications and core technologies have a synergistic impact on this shift towards a customer defined market place. This circumstance is expected to create yet further refinements in customers' perceptions of their needs and requirements. The combination of these forces calls for both differentiation of the existing commodity-type model and extension of the development and domain of future products and services across the meter "interface" point.

The Electric Industry in Transition

INTRODUCTION

Revolutionary change is sweeping the electricity industry on a global scale. Privatization, reregulation, competition, and the electrification needs of developing countries are the synergistic forces which, coupled with convergent advances in technology, are driving business transformation to a level heretofore unparalleled. Indeed, experience gained in markets such as the U.K. indicates that an accelerating pace of change and an ever more complex set of conditions are fuelling future service developments. The only certainty is that products and services within the electricity industry's traditional franchised area of operations will continue to experience rising levels of exposure to competitive pressures resulting in major economic shifts in the industry significantly different from any in the past.

The challenges facing companies in the electricity industry are to exploit this wave of evolving products and services and to seize the resulting opportunities for renewed success and growth. The alternative choice, to either resist or ignore the changes that are occurring, can lead to a path best characterized in a quote by the historian Arnold Toynbee. "In evolution...nothing fails like success. A creature which has become perfectly adapted to its environment, an animal whose whole capacity and vital force is concentrated and expanded in succeeding here and now, has nothing left over with which to respond to any radical change....It can, therefore, beat all competitors in the special field; but equally, should the field change, it must become extinct."[1] Successfully meeting these radical changes requires both a mastery of the forces at work and a willingness to examine the world from a different paradigm. Technology, which is no longer the exclusive realm of engineers or information technology professionals, heads the list as one of the more pervasive drivers of change. For the astute, technology can offer the means to shape the business of tomorrow.

Industry Overview—Snapshots in Time

Past. The vertically-integrated electric utility in combination with tariff-based regulations worked well in the past. Customers wanted "electricity" to operate businesses, tools, homes, appliances, etc. Products offered by utilities to their customers concentrated on producing desired

---

[1] Pascale, Richard, *Managing on the Edge*, The Penguin Group 1991.

changes in the utility's load profile and generation demand curve. Other independent power producers began to emerge in the form of co-generators that competed for major industrial concerns. Small distributed generation and renewable energy applications were limited. Major investments were made largely in generation. Optimization efforts were focused on scheduling generators. Distribution networks were designed to perform passively. Limited penetration of distributed automation technology existed within low voltage networks. Meter "interfaces" were largely electromechanical devices with cumulative monthly registers read manually. Trial initiatives were conducted for automated meter reading from hand-held devices, remote dial-up, and wireless radio technology, but there were only limited interactive communications and no integration between interfaces and network assets. Dynamic integration between the customer's energy requirements and the utility's network operations did not exist.

Present. The industry is being unbundled. Retail wheeling exists. Price capping of regulated service elements is replacing earnings based on return on assets or return on capital. The customers determine from whom they will purchase their electricity. Contract profitability and risk "portfolio" management become key concerns for electricity suppliers. A futures market appears. Asset utilization and stranded capacity become key concerns for utilities. Customers demand base and enhanced services from their suppliers such as next day period-pricing forecasts, demand profiles, aggregate and flexible billing arrangements, energy efficiency consultation, etc. Integrated resource planning among generation, transmission, and distribution increases in complexity. Concerns rise over grid integrity. Dial-up interfaces are installed to manage customer contract and billing information requirements. Distribution automation technology is introduced to improve load management and reduce risks and costs associated with the utility's energy purchases. The meter interface becomes a separate business service offered to customers. Telecommunication and cable companies contract to provide interface services directly with the customer or the electricity supplier. Secondary markets open for interface firms selling information on energy demand profiles to utilities for use in their network planning and management of operations and to energy brokers for use in designing contract offerings. Consortiums, including representatives from telecommunications and cable company subsidiaries, form to broker sales of electricity and gas (energy) directly to the customer.

The Electric Industry in Transition

Future. Technological advances aided by the electrification demands of the developing countries make available to customers low-cost, affordable, distributed generation, renewable energy, and storage capabilities. Customer requirements emerge for remote controlled energy management of these in-house technologies in combination with utility-provided services. A market arises in which customers agree to pay premiums for energy obtained from renewable sources. Ten percent of all new vehicles sold are electric powered. Research in fuel cells and batteries has made major technological breakthroughs, though the full range of resulting applications is yet to be understood. The utility's efforts towards integrated resource management and network operations continue to grow more complex. Power quality becomes a major factor. Interactive communications to support management of customer-owned in-house generation and to ensure integration with industry resources becomes a necessity. Distribution automation efforts are extended to integrate networks with "interfaces" and to introduce "smart" processors for automatic control of operations. Distribution networks are re-engineered to become "smart" networks enabling dynamic management of loads and power requirements. Engineering network designs include utility-owned distributed generation and storage. The sheer volume and complexity of data to be processed, simplified, and adjusted for decision-making purposes continues to climb. Domestic residential customers demand a "one-stop shop" that can offer them the convenience of signing up and paying through a single service for their electricity, water, gas, cable, and telecommunications needs.

Technology as a Change Agent

Technology enables the development and delivery of services and products to increase both business profitability and positioning for new markets. Emerging convergent advances between information technology, telecommunications, and core industry technologies have raised the significance of this business driver to a new level of importance. Today's technology not only facilitates change but can, in and of itself, originate change by actually creating new markets, products, and services. Evidence of this can be seen in results of changes emerging from new generation technologies such as cogeneration and combined-cycle gas turbines. The capabilities developed here provide a basis for creating the retail wheeling market. Continuing technological developments in smaller distributed generation, storage, photovoltaics, etc., can be expected to produce yet other opportunities and greater shifts in the traditional electricity markets.

Chapter 18: Business and Technology Change

Nathan Rosenberg, an economist of technological change at Stanford University, states, "In hindsight, the future was obviously not obvious to contemporaries of most major 20th century inventions." For instance:

- Western Union turned down the chance to buy Alexander Graham Bell's 1876 telephone patent for a mere $100,000. The lack of enthusiasm greeting the telephone was likely rooted in the failure of the human mind to imagine a long-distance communications device that served more purposes than those already served by the telegram.

- IBM only envisioned 10 to 15 orders for computers in 1949. Even by 1956, a Harvard physicist developing the machine said he would be surprised if billing departments ever made use of such a number cruncher. The fact that early machines relied on transistors probably limited the imagination.

- Developers of the technology underlying the video cassette recorder, now widely known as the VCR, thought its commercial market was limited to television stations. Subsequent improvements in design and manufacturing made it saleable to households.

History indicates at least five constraints on human ability to predict the value of new technological developments. Those are:

- The initial primitive understanding of innovations

- The specialized use to that many technological innovations are initially applied

- The complementary and competitive relationships among technologies

- The limited capacity of humans to envision entirely new technological systems rather than simple improvements to existing systems

- The need for technologies to pass economic as well as technological tests of their value

Rosenberg's conclusion is that, "What is called for is not just technical expertise but an exercise of the imagination."[2]

---

[2] Nathan Rosenberg, "Humans Not Great at Assessing New Technology," Stanford News Service, Stanford University, June 3, 1994.

Technology is "acting as a virus, infecting the...economics...and, in turn, driving a complete transformation of the industry. Regulatory change is the consequence, not the cause."[3] Technology has afforded choices to the customers and it is this fact, coupled with the customers' increasing awareness of their own authority, which is shaping tomorrow's markets and constitutes the challenge of today.

Customer Choices and Market Segmentation

"In stark contrast to almost all other industries, applications and services for electric power customers have changed little in the last 30 to 50 years."[4] The business of generating electricity, not the marketing or shaping of the resultant services, has been the principle focus of the utilities serving these markets. This practice is now being contested through (a) core technological advances facilitating the growth of cogeneration facilities and independent power generators, and (b) reduction of generation demand through better energy management practices. As noted in *Forbes'* May 1994 article titled "Another Monopoly Bites the Dust," "...most electricity consumers have been obliged to buy their power from local monopoly suppliers. That is about to change...."[5] The industry as a whole is experiencing a fundamental shift in focus from those products and services set by the utility, designed for rate payers-subscribers-consumers, etc., to a market where the shaping of requirements for services are defined by the customer.

The immediate effects of these changes is to empower customers with freedom of choice in selecting their preferred electricity supplier. The results of the choices being exercised can be observed in the U.K. markets. There, competing for supply of electricity and services to customers is based on a phased defranchising of the industry with competition introduced to the 1+megawatt market (approximately 5,000 customers) in 1990, the 100+kilowatt market (approximately 55,000 customers) in 1994, possibly to the 60+kilowatt market (approximately 350,000 customers) in 1996, and the full market (approximately 23 million customers) in 1998.

---

[3] James M. Piepmeier, David O. Jermain, and Terry L. Egnor, "Breakup of the Bell Monopoly: Lessons for Electric Utilities," The Electric Journal, Vol 6, Number 6, July 1993.

[4] Roger D. Levy, "Competitive Implications of the Information Superhighway to the Electric Power Industry," presented to DA/DSM, January 19, 1994.

[5] "Another Monopoly Bites the Dust," Forbes, May 23, 1994.

Chapter 18: Business and Technology Change

Early results indicated that customers were most likely to choose the supplier offering the lowest unit price of electricity. Experiences now show an increasing sophistication amongst customers in their requirements and expectations for both base and enhanced services. New services designed around information content, such as next day spot market electricity price forecasts and historical energy usage profiles, are expected as standard inclusions in base contract agreements. Product information content is increasing. These enhanced customer requirements necessitate efforts by the utility to acquire more information from across the meter interface point (the interface) to achieve, at a minimum, extensive knowledge about the customers' business operations and their overall energy, not just electricity, demands. The ability to obtain and use this information becomes key to both meeting rising expectations of service requirements from customers and in shaping future products and services for growing market share. Just as importantly, this knowledge is seen as being critical to the electricity company's ability to predict profitability for their products and services and to manage inherent risk of an ever-increasing portfolio of energy purchases and sales contracts. In the absence of these capabilities, economic success is uncertain.

Results of surveys designed to understand customer requirements, such as the one conducted by ICM in 1993 for a utility conference in London (Figure 1), indicate, not surprisingly, that customers' first emphasis is on

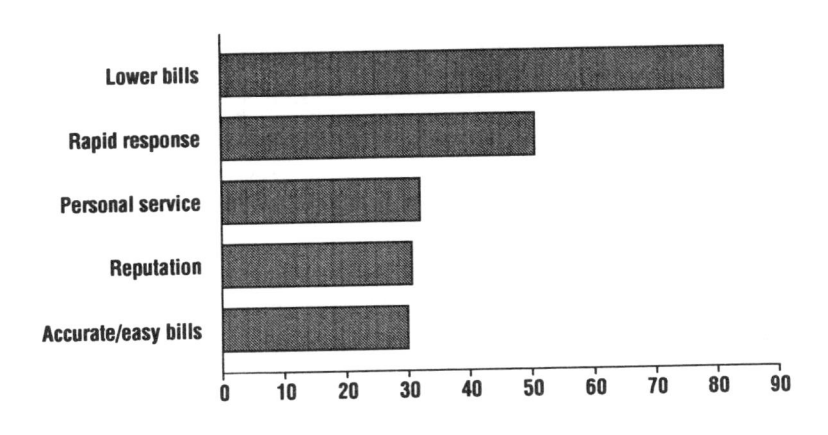

Figure 18-1. Results from a Survey to Understand Utility Customer Requirements

Source: ICM 1993

lower bills. However, this is closely followed by other requirements, such as rapid response by the supplier to the customers' changing needs and the ability of the supplier to personalize service offerings. Again, experience points to the need for information content and significant integration between those business operations occurring on the customer's and utility's side of the meter interface device.

"Giving Customers What They Want" was the theme of the Electricity Supply Conference held in London on 23–24 May. This gathering conducted an early analysis of experiences arising from defranchising the 100+kilowatt market. Table 1 summarizes these findings.

Customers now expect utilities to pay attention to their total energy needs, not just their electricity needs. Towards this end, several electricity suppliers are forming alliances with gas suppliers. It is also worth noting the emergence of innovative opportunities in auxiliary services, such as the provision and management of distributed in-house generation. These services foretell future settings in which management of customer-owned or customer-leased generators and storage will be provided as a service by the utilities. The operation of these facilities will be scheduled in accordance with least-cost demand management principles, like the least-cost routing services being offered today in telecommunications.

Table 18-1. Giving Customers What they Want

Base Services	Differentiators	New Services
Electricity Usage Report	Contract support	Innovation
	• information services	• tailored offer
Energy efficiency	- indicators	• hybrid deals
developments	- analysis	• aggregation of demand
	• innovation	• demand side bidding
Contract offerings	- aggregation	• auxiliary services
	- price	• in-house generation
	- risk	
	- spread	
	- costs	
	• billing	
	- flexible	
	- aggregate	
	- reliable comms.	
	- electronic	
	• customer focus	
	- contract terms	

Source: Electricity Supply Conference, London, 23–24 May

Chapter 18: Business and Technology Change

The increasing sophistication of the customers in defining their wants, combined with a growing emphasis for requiring services containing high information content across the interface, is expected to promote yet further refinements in customers' perceptions and vision of their needs. Additionally, growing understanding of the opportunities afforded by information and technology is expected to continue to broaden the possible range of products and services. In the U.K., note the heightened customer participation in evolving products and services, in contrast with a weaker interest in initiatives such as the demand-side management launched by the utilities. The latter efforts, instead of being driven by customers, are understandably designed more to influence customers' use of electricity in ways to produce desired changes in the utility's load shape.[6]

Experiences gained through the U.K.'s privatization and competition, in combination with emerging information and technological advances, suggest an evolving process for defining market segmentation. This segmentation, illustrated in Figure 18-2, ranges from an area of high-priced sensitivity representing the traditional franchise commodity products and services to a value-added differentiation of these same products, as afforded through either stringent guarantees of services and service level agreements or increasing information contents such as historical energy

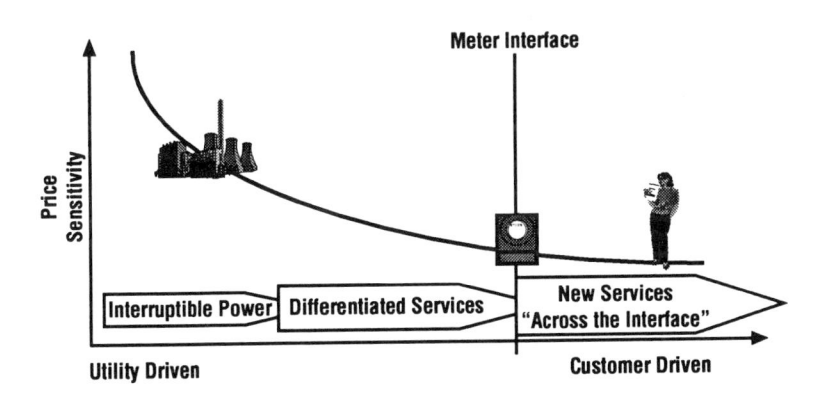

Figure 18-2. Price Sensitivity vs. Products and Services

---

[6] Clark Gellings, and J. Stuart McMenamin, Energy Economist Conference, November 1993 145/21.

demand profiling, followed by a focus on serving the customer's total "energy" requirements (e.g., gas and electricity). Lastly, this segmentation continues through creation of new products and services developed through an intensive information-based component and the integration with technological advances in telecommunications, microelectronics, information technology, and core technology developments in distributed generation and storage, etc.

Future opportunities lie in the development of the markets for these products and services.

Products and Services

To become competitive in these changing markets, utilities need to rethink their business operations and refocus their products and services development programs on being part of a larger information process whose design must provide the knowledge necessary to differentiate commodity-based services, create customized enhanced services, and increase operating efficiencies in core operations to drive down prices and costs. This shift is also fundamental for positioning entry into the new markets expected to arise from the emergence of core and associated technologies to meet future requirements demanded by the customers.

The future environment will see an increase in in-house customer technologies providing electricity generation and storage. Remote management and control of these facilities, their maintenance, and even the utility's purchase of excess capacity for resale and distribution may be expected. As the intelligence embedded in the interface increases, distribution networks will need to shift from passive systems to "smart" infrastructures in order to realize the synergistic benefits of these developments and be positioned for the next wave of product offerings.

An approach to be considered can be illustrated as a simple manufacturing process (Figure 18-3).

Fuels enter the system and are used to create energy. A "black box" then converts these inputs into outputs known as products and services that are delivered to the customer. Envisioning the design and construct of a supporting information process from these possible requirements, one can see where the customers' needs and demand for products and services will be the overriding drivers for all other activities occurring within the "box." Thus knowledge of what is going on behind the interface in the customer's operations and premises becomes essential to planning, developing and

Chapter 18: Business and Technology Change

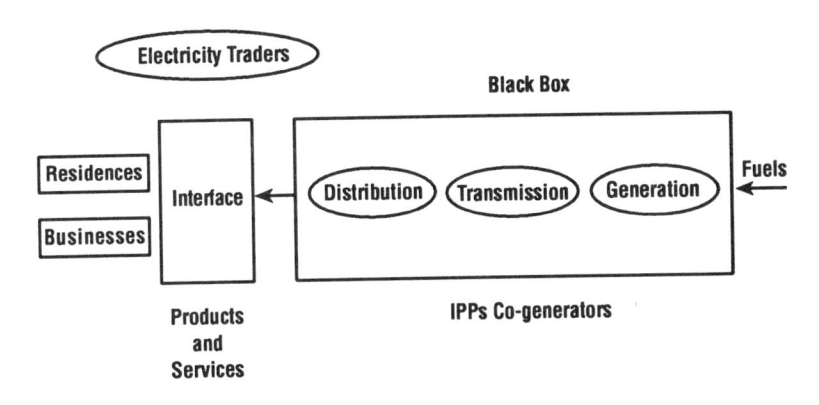

Figure 18-3. Black Box Model for the Electricity Industry

delivering services. Experiences in the U.K.'s privatization activities show that utilities will have to emulate or implement advanced communication and measurement techniques just to meet the basic service requirements of their customers. These investments should be made both with a view of the here and now and considerations as to what competing in future markets may require. "Communications and metering systems make up the largest single program cost. Less capable attempts jeopardize long term operational flexibility."[7] Thinking in this broader perspective is imperative to prevent focusing on niche technologies such as automated meter reading, distribution automation, or demand-side management without understanding their relationship to the whole of the knowledge process.

What is the broader perspective on the markets and consequent knowledge requirements to be considered in building the process? Figure 18-4 provides an illustration of industry and market elements expected to be integrated through a combination of physical connections and virtual information links. Necessary information interfaces with non-utility power generators and electricity suppliers are assumed to be included. The ability to integrate and operate new technology with the traditional infrastructure is critical to the achievement of strategic industry solutions

---

[7] Philip Hanser, Wade Malcolm, and Roger D. Levy, "Re-engineering DSM: Opportunities Through Information and Integration," The Electricity Journal, November 1993, page 27.

The Electric Industry in Transition

Industrial & Commercial Customers

Distributed Generation

"Smart" Interface

INFORMATION DEFINED
- Price forecasts
- Demand profile data
- Customised contracts
* Electronic billing and collections
* Load & energy Consultation
* Tariff negotiation
* Business process redesign

Local Storage

Distribution & Supply
Distribution Substation
Transmission
Meters
Primary Substation   Bulk Supply Point

TECHNOLOGY DEFINED
* Power mgmt. outsourcing
* Building mgmt. services
* Energy management
* Electro technology
* Maintenance

Residential Customers

DIFFERENTIATED SERVICES
• New Supply
• Customer appointments
• Power quality
- Time-of-use pricing
• Change of tenancy
- Automated meter reading
• Loss of supply

Generation

• *Indicates guarantees of service*

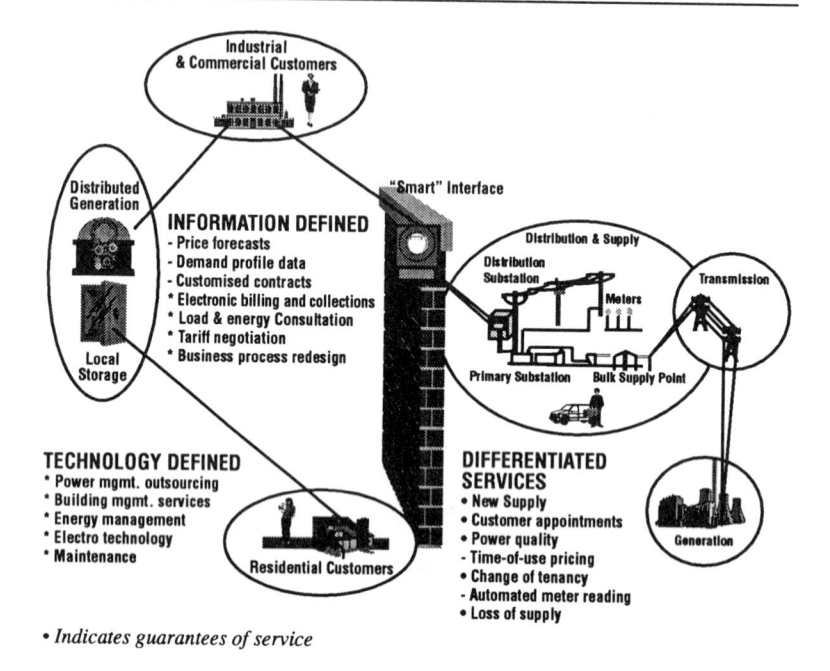

Figure 18-4. Knowledge-based Products and Services

such as integrated resource management and essential to product and service diversity. We need to manage an increasingly diverse and complex resource environment where combined business drivers are creating discontinuous change. On the most basic level, an increasing emphasis on conservation and improvement in asset utilization can be expected. At present it is "estimated that 60% of the cost of electricity is needed to serve peak utility system demand for generation, transmission, and distribution."[8] The drive to more effectively integrate these systems calls for an investment "trend [that] suggests transmission and distribution capital outlays will exceed generation expenditure through the year 2000...Distribution facilities are often poorly utilized...peak loads on many distribution feeders occur for only a few hours per year."[9]

---

[8] Henry Stein, "Utilities are Now Exploring Demand Side Management Uses," Electric Light and Power, January 1993, page 27.

[9] Joseph J. Iannucci, Lynn Coles, Jeremy Bloom, "The Distributed Utility—Is This The Future," 6th International Conference & Exhibition for the Power Generating Industries, November 17–19, 1993, Book IV, pages 33–40.

Chapter 18: Business and Technology Change

Beyond these basic economic considerations are the knowledge requirements for differentiating base services and those for developing new products. Referring to the U.K. model to consider guarantees of service as one service differentiator, we can examine standards for customer appointments that cover special visits to the customer's premises and illustrate a means for market branding. Offerings vary between companies. Some provide morning or noon scheduling windows, others offer next day specific timed appointment schedules backed by a guarantee that, if the service representative is not at the premise within a 15-minute period from the planned visit time, the company will pay the customer a fee of L20 (US $30). The customers of this service are satisfied through improved knowledge of workforce and stock availability, locationing, job stocks matching, etc. Some companies in the U.K. see this as an opportunity to create customer loyalty before the final phased opening of the competitive market.

The challenge of offering either enhanced services or entering new energy-related markets is in constructing an approach that, while acquiring the necessary information, imparts sufficient knowledge to minimize risk and affords additional knowledge, thereby creating a leveraged advantage in optimizing core business operations.

Convergence of Business and Technology

A shift from our current paradigms and the application of our imagination are prerequisites to examining the potential consequences of the business and technological convergences occurring today. Convergent advances in the areas of telecommunications, microelectronics, information and technology, and tv-like user-friendly devices, facilitated by our ability to convert information into digital formats and then compress it by orders of magnitude, will enable development of future applications that cannot be fully comprehended today. Figure 18-5 illustrates a North American model denoting various technologies and a sample of the companies involved in the technological advances bringing the "Information Superhighway."

We can begin to appreciate the potential impact of these advances through the frequent newspaper articles either announcing possible mega-mergers or citing competition between companies in the telecommunications and cable industries for what appears to be each others' core business services. The basis for both the alliance or the competition is a desire to be selected as the chosen provider(s) of "Information Superhighway" services to

The Electric Industry in Transition

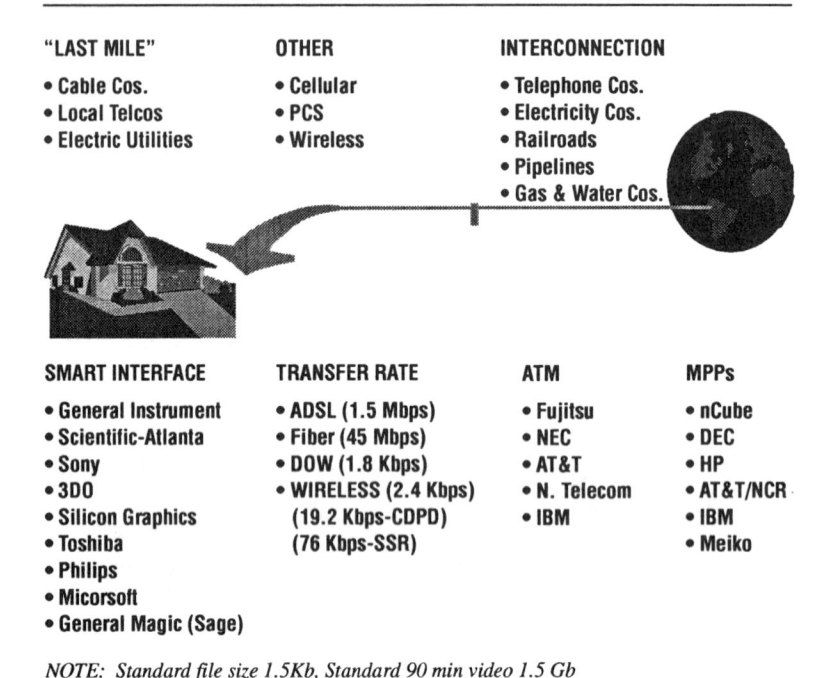

"LAST MILE"	OTHER	INTERCONNECTION
• Cable Cos. • Local Telcos • Electric Utilities	• Cellular • PCS • Wireless	• Telephone Cos. • Electricity Cos. • Railroads • Pipelines • Gas & Water Cos.

SMART INTERFACE	TRANSFER RATE	ATM	MPPs

• General Instrument
• Scientific-Atlanta
• Sony
• 3DO
• Silicon Graphics
• Toshiba
• Philips
• Micorsoft
• General Magic (Sage)
	• ADSL (1.5 Mbps)
	• Fiber (45 Mbps)
	• DOW (1.8 Kbps)
	• WIRELESS (2.4 Kbps)
	  (19.2 Kbps-CDPD)
	  (76 Kbps-SSR)
		• Fujitsu
		• NEC
		• AT&T
		• N. Telecom
		• IBM
			• nCube
			• DEC
			• HP
			• AT&T/NCR
			• IBM
			• Meiko

NOTE: *Standard file size 1.5Kb, Standard 90 min video 1.5 Gb*

Figure 18-5. Convergence of Business and Technology

businesses and homes. Retail shopping, financial services, education, news services, travel, entertainment, etc., are just some of the markets that we understand are awaiting to be transformed by these developments.

In the not too distant past, differentiation in the telecommunications industry was gauged by such minor details as whether one had a private number or party line. Plain old telephone service (POTS) included a phone number, a standard black phone device, and an option to list or hold your assigned number private. Evolution in this industry brought a realization that the differentiating part of the service offering was not in how well the network was engineered or operated but rather in the information-based content of these services. In fact, customers soon adapted a standard for satisfactory performance that assumed the telephone service would be operable at all times regardless of conditions. Telecommunications companies began to offer services such as call waiting, call forwarding, conference calling, voice mail, etc. These new services responded to a growing customer requirement to improve access to information. The offering of these products was made possible by leveraging technological advances in the industry's core processor and switching technologies. A significant driver in these developments was economic. Increasing pressure to lower base phone service rates caused companies to review opportunities where technological advances could create new services whose base marginal cost was comparatively small in relation to marginal revenues. The current emphasis on the "last mile" connection is that this connection is (a) the means for establishing a physical presence at the customers' premises to further ownership of the customer, (b) a barrier to entry for competitors, and (c) a conduit for marketing and delivering future services. Competition between telecommunication and cable companies for this connection is fierce. The stakes are high. The company owning the "last mile" connection and interface owns the right to offer new market services.

What then should these advances and announcements signal to the electricity industry? The situation described above is similar to that faced by the electricity companies today, with one significant exception. Open competition in electricity supply and associated unbundling of the vertical market serve to increase pressures for price reductions of commodity elements. In some spheres of the world, re-regulation includes proposals for price-capped based controls vs. more traditional approaches such as return on assets or capital. What does this mean? In the broadest economic context, holding onto ownership of the "wires" in itself will not secure profitability. Regulatory and competitive scrutiny of transaction components will continue.

What are the potential implications of these convergences in technology to the electricity industry? The interface clearly forms a vital ownership link between the customer and the utility. Interactive communications focusing product and service development on information content are required. Dynamic, not passive, interaction between patterns of customer demand and network operations must exist. Considering the interface as simply an advanced meter technology capable of tracking electricity consumption in shorter, more frequent intervals fails to capitalise on understanding (a) the underlying reasons to these patterns, (b) the potential opportunities for customized services, and (c) what information is necessary to optimize the utility's core business operations to ensure profitability. The interface is the physical manifestation of the utility at the customer's premise. Professor Stephen Littlechild, head of the U.K.'s Office for Electricity Regulation (OFFER), notes that "[competition] will require a

The Electric Industry in Transition

rapid...change in electricity meters, and in the way information is communicated to and from them. Modern developments...offer the prospect of significant benefits for all electricity customers....I believe customers want and deserve something better."[10]

The significant consideration for the electricity companies, mentioned in the previous paragraph, is competition for ownership of the interface. While it is apparent that competition exists between the telecommunications and cable companies for this information conduit, what may not be evident is the potential for competition both against and by the electricity companies for this same connection, even though the business origins of this competition may differ. Specifically, advances in microelectronics and the utilities' need for interactive communications links to the customers may place the utilities in competition with both the telecommunications and the cable companies through efforts to service the interface. Indeed, these interfaces or smart meter solutions can, as the telecommunications connection, use data-over-wire technologies that "possess information-carrying capacity sufficient for use also in supplying services for home shopping, national lottery, road conditions alerts, in addition to expected enhanced services offering control of appliances, etc."[11]

Some would suggest that the "utilities' need for a hardwired connection [to the customer premise] will not go away anytime soon, and that their interest in conserving energy resources can readily justify the investment in an end-to-end fiber optic network that would intrinsically have the capacity to carry a 500 channel television, two-way data communications, video telephony, and more."[12] Conversely, an absence of that knowledge to be acquired from the interface leaves the utility with limited opportunity to shift from a commodity service offering and, denies them the information to refine and re-engineer network operations towards optimization and improved asset utilization.

---

[10] Tim Turnbridge, "Remote Metering Moves Closer," Electrical Review Vol 227 No. 2, January 1994.

[11] John Stansell, "Smart Meters Cut Energy Bills" London Sunday Times, 12 Dec 94, Innovation section.

[12] Ted Bunker, "The Extra Mile," LAN Magazine, June 1994, pages 65–72.

Chapter 18: Business and Technology Change

Towards the Year 2005

The stakes in the game are high and the suspense is surreal. The changes affecting the electricity industry appear irreversible. Now it is the customers who are redefining the markets and identifying new service requirements—their appetites for change can be insatiable. Technology continues its advance, creating new implications and challenges for the industry even while creating new markets for its applications.

Reviewing the pattern of developments experienced in parallel industries may serve to provide some insight into what the future holds. The telecommunications industry poses an interesting comparison as it offers both the advantage of sharing pressures arising from privatization and re-regulation, in addition to having faced major technological advances that have driven new market entries. Following are comparisons between the U.S. telecommunications market...and the electricity industry:

◆ Advances in satellite communications afforded competitors the means to focus on target markets (industrial and large commercial customers) whose usage and cost economics were well suited to pricing strategies challenging the utility's ownership. Parallels can be seen in the emergence of co-generation and generation technologies supporting the independent power producers' entry into the market.

◆ American Telephone and Telegraph's (AT&T's) resistance to change took on several forms. One was through suggesting potential damage to the network from "foreign" interconnection and transmission. Today, concerns over grid integrity are being raised in response to retail wheeling.

◆ Attempts to introduce competition ultimately led to Judge Green's modified final judgement and divestiture of the Bell System. Fully separated subsidiaries were created to participate in non-regulated markets. Efforts to introduce competition and reduce costs have led to unbundling of the industry, with separation between generation, transmission, and distribution and supply. Diversification into associated markets has followed along with positioning for future services to seek enterprises offering higher returns.

◆ Immediately following the telecommunications industry's unbundling, companies redefined their markets as "information markets". Customer requirements for addressing total energy needs led electricity companies

The Electric Industry in Transition

to redefine their markets and form alliances where appropriate to provide "energy" services. For example, the U.K. electricity companies are forming business ventures to market gas.

♦ Telecommunication companies began to fashion high information content products such as 0800 and 0900 calling services. They also introduced enhanced services such as flexible billing, analysis of calling patterns, and customized contracts providing allowances for aggregation of usage. Emerging standards for base services include energy usage analysis and advice on energy efficiency developments. Differentiators are seen at this point as flexible billing and contracts affording aggregation of demand.

♦ Telecommunications companies perceive future differentiation of services to be dependant upon information content and quality. Fierce fighting for rights to provide access to the electronic information highway of the future is in progress. The phone is seen as a smart device offering one means of controlling the gateway to the services to be marketed. Requirements for future services are seen to require enhanced interactive communications with the customer. Advances in microelectronics suggest that the meter of the future will be a "smart interface" offering the means to acquire knowledge and provide services. This device can be configured as an extension of the network, much as the phone is a physical and virtual extension of the telecommunications infrastructure.

♦ Advances in wireless technology are revamping strategies and creating competition. Companies such as AT&T and McCaw offer a means to bypass the telephone lines to the premises, with the added advantage to the customer of mobility. Selective use of wireless radio systems to provide ubiquitous service connections to rural locations is under way. Advances in distribution and storage technologies targeted for customers' in-house use and use by the electricity suppliers offer a means for reshaping the infrastructure of the industry. Vertical static networks are replaced with regionally distributed smart networks managed dynamically to offer enhanced service products at least cost.

While some of the latter developments envisioned for the electricity industry may not yet have materialized, the customer drivers and technology drivers that could make these events possible do exist. The question to be asked is what are the potential consequences of deferring action and not acting on the signals?

Chapter 18: Business and Technology Change

Announcements foretelling possible mergers and acquisitions within the existing U.K. electricity industry structure are already being circulated in the press, just four years since privatization. Who will be the predators and who will be the prey in these endeavors are topics for discussions over after-dinner drinks. These issues take on more significance when considered in conjunction with experiences from other industries, such as those noted below, which are passing through similar degrees of change.

◆ Within the first two years following deregulation of the U.S. Airlines markets, the number of airlines almost doubled. Yet, within the next eight years half of the then-existing airlines and 70 percent of the new competitors had either merged or gone out of business.

◆ Following deregulation of the U.S. banking markets in the 1980s, the rate of bank failures more than doubled.

◆ Within the first three years after deregulation of the U.S. trucking markets in the early 1980s, over 70 companies disappeared.

It is not clear where each electricity company will stand in the future, but the lessons learned can be summarized as points for consideration when deciding upon future strategies.

◆ Technological innovations dramatically expand markets, products, services, and opportunities for competition.

◆ Information-based knowledge allows for the creation of new markets, and the repackaging of existing products.

◆ Knowledge of the customers and their businesses offers significant competitive advantages in developing and delivering products and services.

Appendix I

Biographies of Authors and Moderators

Arnold R. Adler, P.E., a legislative fellow of the American Society of Mechanical Engineers' (ASME) with the New York State Legislative Commission on Science and Technology, retired as general manager of the General Electric Company's (G.E.'s) Turbine Projects Department in 1987. He served in this position for more than a decade, with responsibility for directing and implementing industrial and utility power generation projects in the U.S. and overseas. His prior positions at G.E. were in design engineering and technical management in aircraft engines, air conditioning, and military avionics. As a member of the Legislative Commission staff, he advises various engineering disciplines and assesses technical programs to support legislative policies and initiatives.

Mr. Adler has a Bachelor of Science degree in Mechanical Engineering from City College of New York and a Master of Science degree from the Stevens Institute of Technology. Mr. Adler was a member of the steering committee for the Electric Industry in Transition Conference. The opinions he expressed as session moderator were made in his personal capacity and not as a representative of ASME or the New York State Legislative Commission on Science and Technology.

John A. Anderson, Ph.D., joined the Electricity Consumers Resource Council (ELCON) in 1980, and was named executive director in 1984. ELCON is a group of large industrial consumers of electricity, who collectively operate more than 650 different facilities in most of the 50 states and, in addition, have sizeable operations in numerous foreign countries. They produce a wide range of products—including aluminum, steel, chemicals, petroleum, industrial gases, glass, motor vehicles, electronics, textiles, paper products, and food. Combined, the 22 members consume over four percent of the total electricity in the United States.

Dr. Anderson holds both M.S. and Ph.D. degrees from the University of Florida, with concentrations in public utility economics and industrial organization. He has written and spoken extensively on electricity issues of importance to large industrial firms worldwide.

Anthony F. Armor is the Director of Fossil Power Plants at the Electric Power Research Institute in Palo Alto, California. He directs a research program to improve fossil-fired and combustion-turbine power plant technology for the utility industry. Mr. Armor has worked for EPRI since 1979. He was previously employed at the General Electric Company in Schenectady, New York, in the Large Steam Turbine-Generator Division.

Appendix I

Mr. Armor holds Bachelor of Science and Master of Science degrees from the University of Nottingham, England. He holds eleven U.S. patents for innovations in steam turbine-generator design.

Peter A. Bradford is the chairman of the New York State Public Service Commission, a position he has held since June 1987. He chaired the Maine Public Utilities Commission in 1974 and 1975, and again from 1982 until 1987. He was Maine's Public Advocate in early 1982. He served on the Maine Public Utilities Commission from 1971 to 1977 and was an advisor to Maine Governor Kenneth Curtis from 1968 to 1971, on oil, power and environmental matters. Mr. Bradford served as a member of the U.S. Nuclear Regulatory Commission (NRC) from 1977 until 1982. During his term, the NRC undertook a major upgrading of its regulatory and enforcement processes in the wake of the Three Mile Island accident. Mr. Bradford is the author of *Fragile Structures: A Story of Oil Refineries, National Security and the Coast of Maine* (1975). He has published numerous articles on utility regulation and nuclear power. Mr. Bradford earned his Bachelors degree from Yale University and his law degree from Yale Law School.

James W. Brew, assistant counsel with the New York State Public Service Commission, advises commissioners and senior staff on energy policy, particularly issues that relate to exempt wholesale generators and bulk power markets. He is staff counsel for issues related to integrated resource planning, capacity bidding, DSM, electric transmission, and Clean Air Act compliance. Mr. Brew is a graduate of Georgetown University and the Georgetown University Law Center. He was a member of the steering committee for the Electric Industry in Transition Conference.

Armond Cohen is senior attorney and Energy Project Director for the Conservation Law Foundation (CLF), where he has been employed since 1984. Founded in 1966, CLF is the oldest non-profit environmental law organization in America. Based in Boston, CLF assists utilities, governments, multilateral development banks, and private environmental and energy organizations worldwide in formulating and implementing sustainable energy initiatives.

In the last decade, CLF has been involved in extensive litigation and negotiation with electric and gas utilities to advance sustainable energy initiatives, including energy efficiency and renewable energy, while minimizing system costs. CLF's collaborative work on energy efficiency programs with the New England Electric System garnered CLF the 1992 President's Environmental and Conservation Challenge Award, the nation's highest environmental honor. In 1993, CLF worked with New England Electric to produce the company's strategic plan, NEESPLAN 4, which aims to steadily and significantly reduce power system environmental impacts, while minimizing system costs.

Mr. Cohen joined the CLF staff in 1984 after a clerkship with Judge Harlington Wood, Jr. of the United States Court of Appeals for the 7th Circuit. He holds a law degree from Harvard University and a degree in American history from Brown University. He has written widely on energy and environmental issues.

Mark N. Cooper, Ph.D., is director of research at the Consumer Federation of America and president of Citizens Research, an independent consulting firm. At the Consumer Federation, he has responsibility for energy and telecommunications policy and analysis, as well as internal consulting duties in survey research and economic analysis. As a consultant, Dr. Cooper has provided expert testimony on behalf of People's Counsels and citizen intervenors before public utility commissions on telecommunications and electric utility matters in over three dozen jurisdictions in the U.S. and Canada. Dr. Cooper has also testified on regulatory, antitrust, and public policy issues dealing with the health care, energy and telecommunications industries before Congress, the federal agencies, and the courts.

Dr. Cooper holds a Ph.D. from Yale University and is a former Yale University and Fulbright Fellow. He has published numerous articles in trade and scholarly journals and is the author of two books, *The Transformation of Egypt* (1982); and *Equity and Energy: Rising Energy Prices and the Living Standards of Lower Income Americans* (1983).

William E. Davis became chairman of the board and chief executive officer of Niagara Mohawk Power Corporation, an investor-owned utility providing electricity to over one and a half million customers and natural gas to over a half million customers in upstate New York, on May 1, 1993. With assets of over nine billion dollars and operating revenues of almost four billion dollars per year, Niagara Mohawk is among the thirty largest investor-owned utilities in the United States.

Prior to joining Niagara Mohawk, Mr. Davis was the executive deputy commissioner for the New York State Energy Office. He has also worked for the New York State Department of Commerce and the General Public Utilities Service Corporation in Parsippany, New Jersey. He served in the U.S. Navy as a line officer in Polaris submarines and as an instructor at the U. S. Naval Academy.

Mr. Davis holds a Bachelor of Science degree from the U.S. Naval Academy and a Masters from George Washington University. He is active on a number of boards and committees, including the Business Alliance for a New New York, Canadian Niagara Power, the Center for Clean Air Policy, The Edison Electric Institute, The Energy Association of New York, the Energy and Transportation Task Force of the President's Council on Sustainable Development, HYDRA-CO, the Metropolitan Development Association, the Nuclear Energy Institute, Opinac Energy, Syracuse University and Utilities Mutual Insurance.

Theresa A. Flaim, Ph.D., is currently vice president of corporate strategic planning at Niagara Mohawk Power Corporation. Her department's main responsibilities include integrated electric resource planning, demand side planning and evaluation, incentive regulation, sales forecasting, and strategic planning. Its current emphasis is on assessing emerging competition in the electric utility industry and developing strategies for responding to it, including regulatory and pricing reforms, mitigation of incumbent burdens, restructuring, alternatives for handling transition costs, and improved internal budgeting and planning techniques.

Previously, Dr. Flaim was the manager of the Gas Rates and Integrated Resource Planning Department in Niagara Mohawk's Gas Business Unit, and the director of Electric Demand Side Planning. Prior to joining Niagara Mohawk in 1985, she spent seven years at the Solar Energy Research Institute conducting research related to the economic impact of alternative energy sources on conventional electric utility systems. She received her Ph.D. in Energy Resource Economics from the Agricultural Economics Department at Cornell University in 1977. Dr. Flaim helped plan this conference by participating as a steering committee member.

David Freeman became the president and chief executive officer of the New York Power Authority on March 1, 1994, where he directs operations of the nation's largest non-federal public power organization. The New York Power Authority's eleven power plants and 1,400 circuit miles of transmission lines supply about one-quarter of New York State's electricity.

Mr. Freeman has served in various government positions, including that of energy advisor to former President Jimmy Carter. He is a leading implementer of energy efficiency and of the environmentally compatible production of electricity and has headed the Tennessee Valley Authority, the Lower Colorado River Authority, and the Sacramento Municipal Utility District (SMUD) during the past 15 years. At SMUD, he initiated the nation's most intensive utility conservation program, which is supplying 100 percent of SMUD's growth, and has directed the development of a diversified array of power sources, including renewable ones. At the Lower Colorado River Authority, Mr. Freeman persuaded his Board to abandon plans and equipment for an environmentally damaging lignite mine, and while at the Tennessee Valley Authority, he canceled or halted construction of eight nuclear power plants after determining they weren't needed. He launched a $1 billion air pollution mitigation effort, established an energy efficiency program that reached more than 1 million homes, and made the giant federal agency more responsive to the public.

Mr. Freeman is the author of *Energy: The New Era,* and earned a Bachelor of Science degree in Civil Engineering from Georgia Tech and a law degree from the University of Tennessee Law School. He is a veteran of World War II, during which he served in the U.S. Merchant Marine. He directed the Ford Foundation's landmark Energy Policy Project in the early 1970s and developed the first major study to spell out in detail how energy efficiency could be a major energy resource.

Charles R. Guinn, appointed deputy commissioner for policy and planning of the New York State Energy Office on July 21, 1980, supervises the Division of Policy and Planning that analyzes and evaluates energy policy for the governor and legislature. He chairs the energy planning coordinating committee that develops and updates New York State's Energy Plan.

First chairman of the National Association of State Energy Officials, he was recently appointed chairman of the U.S. Department of Energy's States' Energy Advisory Board. He has participated in many federal and state working groups and numerous National Governors' Association task forces. Mr.Guinn's administrative appointments with the State Energy Office include tenures as assistant deputy commissioner and as chief of the Bureau of Policy Analysis and Planning. In addition he has worked with the New York State Economic Development Board, the New York State Office of Planning Service, the New York State Office

Appendix I

of Planning Coordination, and the New York State Department of Public Works. Before joining the state government he was a transportation planning engineer with the New Haven (CT) Development Agency.

Mr. Guinn has a Bachelor of Science degree in Civil Engineering from Penn State and a Master of Science degree in Civil Engineering from Northwestern University. Mr. Guinn was a member of the steering committee for the Electric Industry in Transition Conference.

Graham H. Hadley is the executive director and managing director of International Business Development for National Power PLC, a privately-owned electric company established in 1990 in the United Kingdom. Between 1983 and 1990, he served as secretary to the Board of the former CEGB (the United Kingdom's government utility), and later he was closely involved in preparations for privatization of the electric utility industry in the U.K. In particular, he played a key role in the design and implementation of the arrangements for splitting up the CEGB and setting up the successor companies.

Previously, Mr. Hadley had held a number of posts within the U.K. Civil Service, including those of under-secretary and head of the Electricity Division of the Department of Energy. Mr. Hadley was educated at Eltham College and Jesus College in Cambridge, from which he obtained a degree in Modern History. In 1991 he completed the Harvard Business School Advanced Management Program.

William W. Hogan, Ph.D., is the Thornton Bradshaw Professor of Public Policy and Management at the John F. Kennedy School of Government, Harvard University. Hogan is research director of the Harvard Electricity Policy Group, which is developing alternative strategies for the transfer to a more competitive electric market. As director of the Project on Economic Reform in Ukraine, he is providing assistance on the broader problems of economic transition to a market economy. He is former chairman of the Public Policy Program at the Kennedy School and a former director of the Energy and Environmental Policy Center there.

He has served on the faculty of Stanford University where he founded the Energy Modeling Forum, and he is a past president of the International Association for Energy Economics (IAEE). He has held positions dealing with energy policy analysis in the Federal Energy Administration, including that of deputy assistant administrator for Data and Analysis. Professor Hogan is a director of Putnam, Hayes & Bartlett, Inc. He is involved in various research and consulting activities including major energy industry restructuring, network pricing and access issues, and privatization in several countries. He received his undergraduate degree from the Air Force Academy and his Ph.D. from UCLA.

B. Jeanine Hull is vice president of Environmental and Regulatory Affairs for LG&E Power Inc., a subsidiary of LG&E Energy Corp. LG&E Power is a fully-integrated company engaged in developing, designing, constructing, financing, owning, operating, and maintaining electric power and cogeneration facilities. Hull is responsible for strategic planning and for environmental and governmental activities related to the energy services industry. Prior to joining LG&E Power in 1988 Hull was project finance counsel at the Washington law firm of Kirkpatrick & Lockhart, and at Lane and Edson, P.C. She also served as counsel to the Energy and Commerce Committee of the U.S. House of Representatives and was responsible for Strategic Planning for the California Energy Commission.

Hull is a frequent speaker and author on a variety of topics involving electricity policy and has testified before Congress on issues ranging from uranium enrichment to transmission access and pricing. She has recently published articles on the reliability of independent power producers, transmission access, and the evolution of the electric industry. In addition, she serves on the Allowance Trading and Electricity Futures Advisory Boards of the New York Mercantile Exchange. Hull holds a Bachelor of Arts degree with honors in Philosophy from the University of Kentucky and a law degree from Hasting College of Law in San Francisco, California.

Leonard S. Hyman is an independent consultant specializing in energy and telecommunications. He was first vice president at Merrill Lynch, Pierce, Fenner and Smith, Inc. where he headed the securities research team on electric utilities, for 17 years. He has participated in privatization work in the United Kingdom, Spain, Mexico, Argentina, Australia, New Zealand, and Brazil. He has served on advisory committees for the U. S. Congress Office of Technology Assessment, NASA, and the Commonwealth of Pennsylvania. Mr. Hyman is author of *America's Electric Utilities,* co-author of *The New Telecommunications Industry,* and a contributor to numerous journals and books. He is presently completing a book on the privatization of public utilities.

Appendix I

Mr. Hyman holds a Bachelors degree from New York University and a Masters degree in Economics from Cornell University. Mr. Hyman helped plan this conference by participating as a steering committee member.

P. Chrisman Iribe, Executive Vice President, U.S. Generating Company (USGen), oversees the nationwide programs for the marketing, development and asset performance of USGen's generating facilities. He has been a principal architect of USGen's strategy for growth since the company's inception in 1989. Mr. Iribe came to U.S. Generating Company in 1989 from ANR Pipeline Company, served as an officer of the American Gas Association in the late 1970's, and has held several federal government positions. Mr. Iribe holds a B.A. in Economics and has completed course work leading to a Doctorate in Economics, both from George Washington University.

US Gen was formed as a general partnership in 1989 by PG&E Enterprises, the non-utility subsidiary of Pacific Gas and Electric Company, and Bechtel Enterprises, the development and financing unit of Bechtel. It provides electricity to utilities and industrial firms nationwide. Based in Bethesda, Maryland, USGen has 19 projects in various stages of development, construction and operation that will have more than 4,000 MW of generating capacity when complete. Of this total, USGen and its affiliates and partners have ownership and management interests in 11 plants, in operation or under construction, totaling 1,712 MW. Together, these facilities represent an investment of $3.7 billion.

Alfred E. Kahn is the Robert Julius Thorne Professor of Political Economy, Emeritus, at Cornell University and a special consultant to National Economic Research Associates (NERA). He has been an advisor on inflation to President Jimmy Carter, chairman of the Council on Wage and Price Stability, chairman of the Civil Aeronautics Board, and chairman of the New York State Public Service Commission. At Cornell, he was successively chairman of the Department of Economics, member of the Board of Trustees, and dean of the College of Arts and Sciences.

Mr. Kahn has been a member of the Attorney General's National Committee to Study the Antitrust Laws, the senior staff of the President's Council of Economic Advisors, the Economic Advisory Council of AT&T, the National Academy of Sciences Advisory Review Committee on Sulfur Dioxide Emissions, the Environmental Advisory Committee of the Federal Energy Administration, the Public Advisory Board of the Electric Power Research Institute, the Board of Directors of the New York State Energy Research and Development Authority, the Executive Committee of the National Association of Regulatory Utility Commissioners, the National Commission for the Review of Antitrust Laws and Procedures, the National Governing Board of Common Cause, Governor Cuomo's Fact-Finding Panel on Long Island Lighting Company's Nuclear Power Plant at Shoreham, Long Island (in 1983) and the Governor's Advisory Committee on Public Power for Long Island, the New York State Telecommunications Exchange, the Ohio Blue Ribbon Panel on Telecommunications Regulations, and a panel of the International Institute for Applied Systems Analysis (IIASA), to advise Eastern European economists and public officials. He is a member of *The American Heritage Dictionary* Usage Panel, and a regular commentator on the Public Broadcasting System (PBS) program, *The Nightly Business Report.*

Mr. Kahn has six honorary degrees, is a fellow of the American Academy of Arts and Sciences, and is a former vice president of the American Economics Association. He has written four books, including the two-volume *The Economics of Regulation,* and some 100 professional articles.

Edward P. Kahn, Ph.D., economist, is leader of the Utility Policy and Planning Group at Lawrence Berkeley Laboratory in Berkeley, California, where he has been employed since 1975. Concurrently, he is employed as a research economist and lecturer by the University of California, Berkeley, with the university-wide Energy Research Group and the Energy and Resources Program.

Dr. Kahn is a nationally recognized expert on electricity, integrated resource planning and regulatory issues, and has counselled utilities, state agencies, and private companies. He has also provided expert witness testimony before regulatory commissions in California, the District of Columbia, Illinois, Indiana, Massachusetts, New Jersey, and Pennsylvania. He is the author of *Electric Utility Planning and Regulation* (1991), and has published articles in *Journal of Regulatory Economics, The Electricity Journal, IEEE Transactions on Power Systems, Utilities Policy, Journal of Political Economy, Public Utilities Fortnightly, Energy Systems and Policy, Energy: The International Journal, Energy Economics,* and *Energy Policy.*

Dr. Kahn received a Bachelor of Arts degree in English from Amherst College, and a Masters of Arts degree in Mathematics and a Doctorate in English from the University of California, Berkeley.

Appendix I

Michael J. Kelleher, director of economic research and forecasting at Niagara Mohawk Power Corporation, directs sales and load forecasting, program evaluation, end-use modeling, and regulatory economics. As manager of resource economics and program evaluation at Niagara Mohawk, he directed DSM program evaluation and was directly involved in integrated resources planning efforts. Mr. Kelleher has a Bachelor of Science degree in Agricultural Engineering Technology and a Master of Science degree in Applied Economics from Cornell University. Mr. Kelleher was a member of the steering committee for the conference on the electric industry in transition.

Frank Kreith, Doc.Sc., P.E., currently serves as the American Society of Mechanical Engineers legislative fellow for energy and environment at the National Conference of State Legislatures (NCSL). In this capacity, he provides assistance on energy management, waste disposal, and environmental protection to legislators in all fifty state governments. Prior to joining NCSL in 1988, Dr. Kreith was chief of Thermal Research at the Solar Energy Research Institute (SERI), currently the National Renewable Energy Laboratory. During his tenure at SERI, he participated in the Presidential Domestic Energy Review, served as one of the energy advisors to the Governor of Colorado, and was the technical editor of the ASME *Journal of Solar Energy Engineering.* From 1951 to 1977, Dr. Kreith taught at the University of California, Lehigh University, and the University of Colorado. He is the author of textbooks on heat transfer, nuclear power, solar energy, and energy management. He is the recipient of the Charles Greeley Abbot Award from ASES, the Max Jakob Award from ASME-AIChE. He received the Ralph Coats Roe Medal from ASME for providing information to legislators about energy and conservation and thereby contributing to the public's appreciation of engineers' worth to society.

Dr. Kreith has served as consultant and advisor on energy planning all over the world. His assignments have included consultancies to the U.S. Department of Energy, NATO, the U.S. Agency for International Development, and the United Nations.

William J. LeBlanc, a project director for Barakat & Chamberlin, Inc., specializes in DSM, strategic marketing, bidding, and pricing for utility clients. He has more than a decade of experience in the utility industry. Before joining Barakat & Chamberlin, Mr. LeBlanc was project manager at the Electric Power Research Institute (EPRI), where he managed DSM projects, marketing, bidding, backup generation, and cogeneration.

As project manager of Demand-Side Management and Rates for Pacific Gas and Electric Company, he managed DSM projects for commercial and agricultural markets. He was responsible for all aspects of the projects, including market research, marketing, program design, implementation, and evaluation.

Mr. LeBlanc has Bachelor of Science and Master of Science degrees in Mechanical Engineering from Stanford University and a Bachelor of Arts degree in Management Engineering from Claremont McKenna College. He founded the Association of Demand-Side Management Professionals (ADSMP), one of the industry's major professional organizations. Mr. LeBlanc was a member of the steering committee for the Electric Industry in Transition Conference.

Francis J. Murray, Jr., was appointed New York State Energy Commissioner by Governor Mario M. Cuomo in June 1992 and was confirmed by the state senate in July. Mr. Murray heads the State Energy Office and is chairman of the State Energy Research and Development Authority. He also serves as chairman of the State Energy Planning Board, a multi-agency board charged with the responsibility for developing a comprehensive integrated energy plan for the state.

Prior to his appointment, Mr. Murray was Governor Mario Cuomo's Deputy Secretary for Energy and the Environment, as which he served as the governor's principal policy advisor on energy and environmental issues. Mr. Murray had been Governor Cuomo's Assistant Secretary for Energy and the Environment from January 1983 until October 1985, when he was named Deputy Secretary. He was responsible for overseeing the development and implementation of the governor's legislative and administrative initiatives in the areas of energy, environment, parks, and recreation. In that capacity, he represented New York in numerous national and regional energy and environmental activities, including the Coalition of Northeastern Governors, the National Governors' Association, and the Council of Great Lakes Governors. From July 1980 until 1983, Mr. Murray served as a staff adviser to Governor Hugh L. Carey on energy and environmental matters. He was a legislative counsel in New York State's Office of Federal Affairs from March 1997 to July 1980, where he served as chief liaison between the governor's office and the New York congressional delegation in representing and advocating New York's interest in energy, environmental, and recreation legislation.

A graduate of Georgetown University's School of Foreign Service and Georgetown Law Center, he served as U.S. representative James V. Stanton's legislative counsel before joining New York's Office of Federal Affairs. Mr. Murray is also a member of the Virginia State Bar Association, the Hudson River Greenway Communities Council, and the New York State Urban Cultural Park Advisory Committee. In October 1994, Mr. Murray was named by President Clinton to the Advisory Committee to Assist in the Development of Measures to Significantly Reduce Greenhouse Gas Emissions from Personal Vehicles, which is charged with developing strategies to reduce greenhouse gas emissions from cars and light trucks.

Philip R. O'Connor, Ph.D., is managing director of Palmer Bellevue, a division of Coopers & Lybrand. Palmer Bellevue provides consulting, financial and demand-side management services to the utility, natural gas, telecommunications, independent power, and related industries.

Dr. O'Connor is a nationally recognized expert on the development and implementation of competitive strategies in regulated industries. He is a frequent speaker, both nationally and internationally, on utility issues and has written a number of articles in professional trade journals. He has served as strategic advisor and expert witness regarding regulation and competitive positioning for New Jersey Natural Gas, United Water Resources, Wisconsin Bell, Michigan Bell, and Indiana Bell. He also served as regulatory and strategic advisor and expert witness to Public Service Company of New Mexico on a proposed corporate restructuring and lead advisor to the Sacramento Municipal Utility in a program to divest ownership and operation of the Rancho Seco Nuclear Generating Station. Prior to forming Palmer Bellevue in 1985, Dr. O'Connor served as Illinois' chief utility regulator, chairing the Illinois Commerce Commission.

Dr. O'Connor received a Bachelor of Arts degree from Loyola University of Chicago, and Masters and Doctoral degrees in Political Science from Northwestern University.

Edward M. Smith is chief executive officer of Perot Systems Europe (Energy Services) Ltd. He is responsible for Perot's energy sector operations throughout Europe and the development of related industry strategies globally. Mr. Smith also serves as managing director and member of the Board for Metricom-U.K., a joint venture created to establish a mobile data communications network and services across the U.K. Prior to joining Perot in 1990, he was director of operations for a private engineering firm focusing on international and U.S.-based utility (telecommunications, electricity, gas, and water) engineering services. From 1979 to 1986, Mr. Smith served as a senior manager with American Telephone and Telegraph's (AT&T's) operations and engineering departments in a variety of leadership roles focussing upon the design and delivery of both public switched and customized telecommunication networks. His experiences before joining AT&T included a brief assignment with Pfizer Laboratories and service as an officer in the U.S. Marine Corps.

Mr. Smith holds a Bachelor of Science degree from the U.S. Naval Academy at Annapolis, an MS degree in Systems Management from the University of South California, and an MBA degree (Executive Program) from Emory University. Mr. Smith is married, with three children, and resides in London, England, with his family.

Charles Stalon, Ph.D., is a consultant on energy regulation. He recently retired from the positions of director of the Institute of Public Utilities and professor of economics at Michigan State University. His principal area of research and writing is the economic regulation of U.S. energy industries. Prior to joining Michigan State University, Dr. Stalon served as commissioner for the Federal Energy Regulatory Commission. Before that, he served as commissioner for the Illinois Commerce Commission and as professor of economics at Southern Illinois University at Carbondale.

Dr. Stalon currently serves as a member of (1) the Board of Directors of New Jersey Resources Corporation; (2) the Advisory Council of the Gas Research Institute; (3) the Advisory Council of Bellcore; (4) the Keystone Energy Advisory Board; and (5) the editorial boards of the *Energy Report,* and *Natural Gas.* In addition, he is an active participant in the Harvard Electric Policy Group.

Dr. Stalon received a Bachelor of Arts degree in Economics from Butler University and Master of Science and Ph.D. degrees from Purdue University.

Marsha L. Walton, Associate Project Manager, New York State Energy Research and Development Authority, oversees the Integrated Resource Planning subprogram. She manages statewide research projects that

Appendix I

investigate environmental externalities associated with electric generation, evaluate DSM potential and utility programs, and support policy research for the New York State energy planning process.

Before joining the Energy Authority in 1992, she was an energy-efficiency analyst with the Office of Energy Efficiency and Environment (OEEE), New York State Department of Public Service, where she had oversight of the New York investor-owned electric utilities' DSM and IRP plans and DSM evaluations.

Ms. Walton holds a Bachelors of Arts degree from Bard College and a Masters degree in Regional Planning from Cornell University.

Ms. Walton planned the Electric Industry in Transition Conference.